Mining Among the Clouds

by Harvey N. Gardiner

Colorado History
ISSN 1091-7438

Number 6
2002

COLORADO HISTORICAL SOCIETY

Research and Publications Office
Modupe Labode and David N. Wetzel, *directors*

Publications Director
David N. Wetzel

Colorado History Series Editor
Larry Borowsky

Cover Design
Mary H. Junda

The Colorado Historical Society publishes *Colorado History* to provide a flexible scholarly forum for well-written, documented manuscripts on the history of Colorado and the Rocky Mountain West. Its twofold structure is designed to accommodate article-length manuscripts in the traditional journal style and longer, book-length works which appear as monographs within the series. Monographs and special thematic issues are individually indexed; other volumes are indexed every five years. The publications of the Society generally follow the principles and conventions of the *Chicago Manual of Style,* and an author's guide is available on request. Manuscripts and letters should be addressed to: Research and Publications Office, Colorado Historical Society, 1300 Broadway, Denver CO 80203. The Society disclaims responsibility for statements of fact or opinion made by contributors.

In memory of my father, C. Harvey Gardiner (1913-2000), professor of history

Contents

Introduction   viii

Chapter 1
Colorado's Reputation for Extravagant, Prodigal,
and Unscientific Metallurgical Experiments   1

Chapter 2
Two Rich Discoveries: The Dwight and the Moose   4

Chapter 3
Organizing the Moose Mining Company   16

Chapter 4
Opportunity Knocks: The Start of a Mining Boom   20

Chapter 5
The Boom in 1872   25

Chapter 6
Silver Replacement Deposits and the Mining Law   30

Chapter 7
Creating an Ore Market   39

Chapter 8
1873: A Good Year Ends on a Bad Note   44

Chapter 9
The Problems with the Ore Market   49

Chapter 10
One Fabulous Mine, One Great Mine, and a
Struggling Ore Market   56

Chapter 11
The Discovery of Leadville, and How It Whipsawed
Park County   63

Chapter 12
The Moose Mining Company: A Favorite of
the Mining Stock Crowd   70

Chapter 13
The Hall and Brunk Silver Mining Company   82

Chapter 14
Living at These High Mines   91

Chapter 15
The Power of the Metals Market   100

Epilogue   105

Endnotes   106

Bibliography   125

Acknowledgments

My research into the Moose Mine received impetus from a Colorado State Historical Fund mini-grant. I wish to thank Carol Davis of the South Park Historical Foundation in Fairplay for administering the mini-grant. With these moneys, I acquired 41 patent case files from the National Archives, title searched 133 deeds in the Park County Records, and had copies made of relevant photographs. I particularly want to thank Adam Bauer of Pikes Peak Title Service in Woodland Park for his prompt title searching.

Many libraries and government agencies assisted my research: the BLM Colorado State Office, where Andy Senti taught me how to research the records; Denver Public Library, Western History Department; Colorado Historical Society Library; University of Colorado Libraries; Colorado Division of Mines; U.S.Geological Service Library in the Denver Federal Center; National Archives in the Denver Federal Center; Colorado State Archives; Kansas City Public Library; Bentley Historical Library at the University of Michigan; and the Amon Carter Museum.

I specifically wish to thank the Interlibrary Loan Department in Norlin Library at the University of Colorado for tracking down numerous materials, some of which were difficult to locate. Sincere thanks to Maury Reiber, owner of the Russia Mine and many other Mount Lincoln claims, who met with me on two occasions, read the manuscript, and offered valuable clarifications. Marvin Moore, former owner of the Morningside-Creskill Group on Mount Bross, and Erik Swanson of Alma also provided information.

Finally, thanks to my wife and very own copyeditor, Phyllis Hunt, for numerous revisions to the manuscript. She was instrumental in turning a complicated mass of research into a readable story.

Introduction

The 1870s were a crucial and transitional decade for Colorado. The territory became a state in 1876, the population grew by 387 percent, and mining recovered from the failures of the 1860s to reign once again as the state's bread-and-butter industry. Moreover, the brief heyday of Colorado's gold mining gave way to silver bonanzas in the 1870s, a major shift in the mining industry. Gold production increased by only 6 percent as silver production soared a stunning 1,900 percent. Where silver had accounted for only 17 percent of the minerals produced in 1870, it represented 72 percent of those produced in 1879. By the end of the 1870s Colorado led the United States in silver production and was firmly fixed in the nation's mind as the holder of hidden bonanzas.

The discovery that fueled this increase in silver production occurred in July 1871 on Mount Bross, in the northwest corner of Park County, when two experienced and patient prospectors discovered silver ore deposited in limestone. Up to that time silver had only been known to appear in granite, and the geological distinction between silver deposits in limestone and silver fissure veins in granite would open a whole new chapter in Colorado mining history. This discovery of bonanza silver ore in limestone at what became the Moose Mine would lay the foundation for Colorado's Silver Decade. It stands as one of Colorado's signal mineral discoveries and makes the Moose arguably the first great silver mine in the state. Incredibly rich, the Moose produced one bonanza strike after another, setting off a mining boom that led Park County's silver production to increase by a staggering 2,455 percent between 1871 and 1875.

The recurring themes of isolation, transportation, and limited ore markets that were reported in mining camps throughout the West also expressed themselves on Mount Bross and neighboring Mount Lincoln. The miners struggled with isolation and poor transportation, which made living and mining expensive propositions. The local mining economy of Park County, when confronted with a nonexistent ore market, attempted to create its own market by building smelters.

Some unique themes also stand out. First, the prospectors and miners had to understand the geology of silver ore in a new form: horizontal replacement deposits in limestone. Second, neither the Mining Law of 1866 nor the Mining Law of 1872 anticipated horizontal deposits. This situation thus led to confusion about what to call these mineral discoveries and how to meet the requirements of the mining law.

Third, the ultimate challenge on these mountains was the elevation of the mines. Located as high as 14,157 feet, these were the highest mines—"aerial" mines as one newspaper account referred to them—ever worked in North America. The physical and psychological challenges of living and mining at such high altitudes included supplying the mines, hauling machinery up to them and ore down from them, and building shelter and livable conditions. Making matters even more challenging, several of the mines, including the Moose at 13,700 feet, operated year-round.

Of monumental importance to the history of Colorado mining is the fact that deposition of silver ore in limestone at the Moose in Park County, on the eastern side of the Mosquito Range, is in the same geological formation as the silver ore later found at the Rock and Iron mines in Lake County, on the western side of the range. The story of the Moose Mine traces how the mining boom that followed the Moose discovery ultimately launched the Leadville boom, which did more for populating the mountains and plains of Colorado than any other single event.

The silver mining boom on Mount Bross and Mount Lincoln was small and short-lived in comparison with Leadville's boom. In the wild, heady times that turned the focus to Leadville's bonanzas and made Colorado the most prosperous state in the Union, the role that Mounts Bross and Lincoln played was quickly forgotten, and its significance left untold. The story of the Moose Mine documents the confusion, activities, and events that led to the Leadville boom, and makes it possible to put the discovery of Leadville into context for the first time.

Mining Among
the Clouds

Chapter One

Colorado's Reputation for Extravagant, Prodigal, and Unscientific Metallurgical Experiments

Every one of the early mining frenzies in the West centered on discoveries of "placer" gold: gold nuggets found in streambeds. With gold seekers appearing in droves and rapidly shoveling up these rich placer deposits, Colorado mining faced some hard lessons about geology and metallurgy by the early 1860s. And many wealthy easterners recklessly invested large sums of money in Colorado mining only to, in the end, lose millions in a faltering mining technology that robbed Colorado of its reputation.

The rich beds of placer gold came from surface ore outcrops that had been exposed to weathering by air, temperature, and water. In geologic time these outcrops, along with the ore near the surface, concentrated and separated from the surrounding rock. Freezing and thawing broke the surface minerals free, and with time and erosion, water buried this placer gold deep in streambeds.

These placer gold discoveries led people into new areas. The prospector with his gold pan and burro, that archetypal romantic image of the West, washed the sands and gravel from streambeds looking for "color." Behind him came more prospectors, then merchants selling food and supplies, and then every other type of financial activity the local economy could support.

Placer mining operated not in geologic time but in human time: hurried and frantic. Beginning in 1859 placer gold discoveries caused a number of boom camps to spring up in Colorado, but in the rush to get the gold the rich placer deposits were exhausted soon after their discovery. Only for a short time could placer gold constitute the primary mineral produced in a mining economy. Then the early prospectors and miners had to confront the fact that recovering the rest of the minerals in Colorado's mountains would demand tedious work. Instead of the easily obtained, quick wealth they had imagined back home, they now had to search for the lode—the *source* of the gold.

As the rich placer deposits disappeared, the pioneer mining men began to locate the surface outcrops of the gold lodes. This surface gold, called free-milling gold, would prove to stamp and crush easily in a mill, as the enclosing rock broke apart like a nutshell to reveal the gold kernel. Mercury could then be combined with the crushed ore to absorb the gold at the same time it excluded foreign matter. The mercury and gold formed a more or less soft, spongy mass, and when the mercury vaporized, the gold was left behind. Mining these surface deposits was, like placer mining, also easy, cheap, and profitable, but only to an extent.

At this point in early-day Colorado gold mining, a basic law of nature entered the story, causing miners great confusion and consternation. Air, water, and weather can only concentrate and separate elements in the earth—in this case, the free-milling gold—to the depth of the water table. Once they ventured below the water table, the miners' problems began. First, water collected in the mines as they went deeper. Next, as the mining passed below the rich free-milling ore near the surface, it went into what was called the "cap."[1] The mineralogy of the gold ores now became "refractory ores," a complex sulfide containing gold but locked in a solid mineral combination with sulfur. Crushing these ores in a mill did not free the gold, and amalgamation with mercury did not attract the gold either. The mineralogy of the cap stumped the metallurgical technology of the time. Where the recovery rate from surface gold ores had been about 75 percent of the assay value, the recovery rate from refractory ores fell to a miserable 25 percent of the assay value. Technological inefficiency on this scale translated into zero profit.

By the end of 1863 this situation started a severe economic contraction in Colorado. Mining companies needed working capital to mine ore. At the same time, the companies needed a metallurgical solution to the gold recovery problems so that they could produce a profit to use as working capital. This sent mine owners east looking for investment money.

A unique economic twist of circumstances worked to their advantage in December 1861, when the Lincoln administration suspended the gold standard. A legal-tender bill authorizing paper money, greenbacks, followed to save the national economy and provide financing for the Civil War. Once specie payments in gold ended, gold became a marketable commodity, and a speculative market for gold began in New York. This market represented the difference between the valuation of gold coin and paper currency. Gold opened in January 1862 at $101 per ounce and, with the ups and downs of the market, rose to $285 per ounce on July 11, 1864.[2] Gold provided protection against changes in the value of the greenbacks, and many wealthy easterners, fearing wartime inflation, concluded that their money was safer invested in Colorado gold mining than in depreciating paper money.

This fueled an investment boom in Colorado mining, but it came before anyone had solved the metallurgical and technological problems of recovering gold from sulfide ores. Nonetheless, newly organized eastern mining companies made plans to reap fortunes. In the summer of 1864 shipping agents consigned 1,000,000 pounds of mining machinery to Colorado. Freighters had to haul everything across the plains, and goods valued at $5,000 at the Missouri River cost $10,000 delivered to Denver. Moreover, Cheyenne and Arapaho warriors had virtually severed the transportation routes.[3]

Doomed to failure, eastern mining companies bought mines and built costly and useless mills that could not efficiently recover the gold. Snared by their desperation, gullibility, and greed, credulous mine owners and investors began purchasing quack inventions reputed to recover gold. Every one failed dismally. Mismanagement and dishonesty ran rampant. One figure put the sum invested by New Yorkers alone at $48,480,000. Of this sum, $2,250,000 built unsuccessful stamp mills. Another $2,000,000 went for hopeless schemes and futile gold-recovering processes. Rusting machinery and assorted contraptions littered mining areas. The gold boom crashed amid near universal financial failures. For years thereafter, many investors expressed complete disbelief in the value or permanency of Colorado mines.[4]

The years 1864–68 represented the lowest point for Colorado mining, in an almost unbroken line of failures that had begun in 1860. It was at this critical juncture that Nathaniel P. Hill (1832–1901), a professor of chemistry and geology at Brown University, became interested in the problem of ore reduction. Hill twice visited the world-famous smelting center at Swansea, Wales, to study the metallurgical processes used there. Upon his return to Colorado he organized the Boston and Colorado Smelting Company in 1867, which built the Boston and Colorado Smelter at Black Hawk. When his smelter fired up in February 1868, it revived Colorado gold mining: Hill had found the right process to recover gold from sulfide ores. But while the future brightened, Colorado still suffered from its reputation for mistakes. Millions of dollars had been spent for wild, optimistic claims about the value of mineral lodes and for unnecessary machinery and processes. As the 1860s ended, Colorado seemed to lack the elements necessary to become a prosperous state, and nothing in Colorado mining suggested that the 1870s would begin gloriously.

Chapter Two

Two Rich Discoveries: The Dwight and the Moose

A look at present-day Park County in the 1860s illustrates the shift in Colorado from gold to silver and the roller-coaster ride of boom, then bust. Thousands of young men who rushed to the Mosquito Range following the discovery of placer gold were counted in the 1860 census, which listed 10,600 people in the broad and comparatively level basin of South Park and the mountains surrounding it. This area became Park County, and it constituted 31 percent of Colorado Territory's total population of 34,277 residents in 1860.

Three mining camps tell the story of the boom and bust. The camp called Buckskin Joe, at 10,500 feet, ran fast and furious in 1860, sporting a theatre, bank, bowling alley, a half-mile of saloons and faro shops, and a large gambling tent. Several thousand miners shoveled the placers or worked underground in the Phillips Mine, the first gold lode found in Park County. Buckskin Joe produced $1,600,000 in gold over the decade and boasted nine mills to crush gold ores. But a visitor to Buckskin Joe in August 1870 wrote: "Fate has written the destiny of this place on these crumbling and dilapidated walls. This was once one of the fastest places in the west, but the people lost their grip and let it decay."[1] Only one mill still operated, erratically, and just three families constituted the population of Buckskin Joe, a near ghost town.

A second camp, Montgomery, located at the base of Mount Lincoln and Hoosier Pass (and now covered by the Montgomery Reservoir), suffered the same hard times. Two thousand people once lived there, the echoes of the dropping stamps in six gold mills giving evidence that the nearby mines were "in pay." Men had speculated in corner town lots. Montgomery produced $500,000 in gold, but one or two years of steady mining exhausted the supply of surface ores. In August 1869, about seventy houses and dilapidated

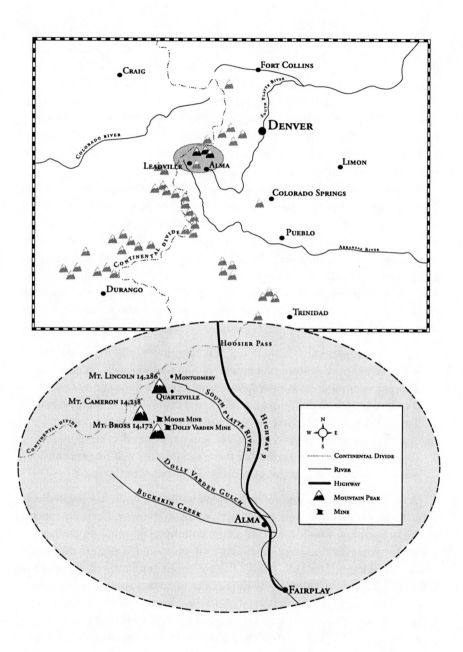

Park County mining district, 1870s. Map design by Mary Junda.

cabins with mud roofs still stood.[2] The population was less than a dozen and included only one family, that of Joseph H. Myers. The 1870 census counted thirty-seven people in Montgomery, with only one mill, the Pioneer Mill, operating. A third camp nearby, Mosquito, had supplied six gold mills. By 1870 none of the mills operated, and Mosquito was home to seven men and one family. In contrast to the 1860 census, which counted 10,600 people in Park County (representing 31 percent of the population in the territory), the 1870 census listed only 447 residents in the county, or 1 percent of the territory's population.

Everyone who could get away from Park County had done so in the floundering, desperate years of the late 1860s. The population of the entire territory increased by only 5,587 between 1860 and 1870, while Denver's population increased by only thirty people. In Park County some placer mining operations continued to work, but the twenty-one abandoned gold mills stood as testament to Colorado's reputation for extravagant, prodigal, and unscientific metallurgical experiments. One blunt description of the cause of Park County's mining problems also summarized the stigma of mining everywhere in Colorado:

> It is not the mines that have failed, it is the people, and their reckless and extravagant system of mining. A more economical process of treating ores must be adopted; one hundred, two hundred and three hundred thousand dollar quartz mills is not the system for these mines. Let men understand these things, and be governed and act accordingly. These veins of gold and silver and copper and other precious metals that so largely abound in these mountains, will be profitably mined when the last of this generation sleep beneath the sod.[3]

As the old mining towns above Fairplay withered, Buckskin Joe ceded its position as Park County seat to Fairplay, eight miles away. Business concentrated in Fairplay, which subsisted as the commercial, political, and supply center for what little economic activity took place on the eastern side of the Mosquito Range. The U.S. Land Office established its third Territory of Colorado branch at Fairplay in 1867. Like residents in all mining towns, Fairplay's residents paid more attention to the glittering sands of gold than to their town, which was a dusty, bedraggled collection of thirty or forty log structures in 1869. These were difficult, slow economic times in Park County. Mine owners lacked capital and seemed little interested in trying to work their properties. An abundance of lodes could be bought for a "song," while other owners would give half interest in exchange for enough money to start doing some development work. At least the South Platte River offered the "Fairplay trout diet": trout for breakfast, trout for lunch, and trout for dinner.[4]

Everyone hoped for better times. The prospectors and miners who remained were, in general, patient of toil and hopeful of success, believing that at some point the mineral wealth of the Mosquito Range would be revealed. Two experienced prospectors, Daniel Plummer, who had come to Montgomery in July 1865 to superintend the Pioneer Mill, and Joseph H. Myers, who had been mining in Park County since the early 1860s, doggedly continued to search for promising mineral outcrops. They were at last rewarded when in 1868 they found an outcrop of silver at 13,600 feet on Mount Bross.[5] Earlier, silver had been discovered at the 10-40 lode near Buckskin Joe, but nothing had come of it. In contrast to the silver found in the 10-40 lode and everywhere else in Colorado, the silver Plummer and Myers discovered on Mount Bross outcropped in limestone—not as a fissure vein in granite—marking it as the first discovery of silver in limestone found anywhere in Colorado.

Unaware of the geological significance of this silver-in-limestone outcrop, Plummer and Myers assumed it led to a fissure vein. They guessed, or hoped, that they could locate other fissure veins down the northeast face of Mount Bross, and that these veins in the aggregate would be from the same lode. Proving this supposition would be hard, however, for the slope here is very steep with slide rock and debris covering the surface. While limestone ledges do appear at intervals, this type of terrain makes prospecting arduous work. Thus they could only guess at the possible configuration of the ore body when they marked the boundaries of their mining claim.

Marking the boundaries on land in the public domain meant entering the realm of mining rules and regulations. Establishing the boundaries of a mining claim on public lands was then predicated upon the rules, customs, and laws that had evolved between 1849 and 1866 to protect the rights of the discoverer. No federal or territorial laws were on the books to govern mining in the public domain before 1866. Technically, all miners on the public domain in the West were trespassers as they organized themselves into mining districts. Lacking even common-law principles concerning mining and mining rights to guide them, miners had to establish order and protect their claims through collective action. They evolved customs based on elementary and well-established rules of property and equity. While specifics of the regulations varied from one mining district to another, they all defined the size of claims, the proper method of marking a claim's boundaries, and the work the discoverer must perform on his claim in order to maintain his right to the mineral. In our nation's history, this collective action by western miners is regarded as direct democracy and popular sovereignty at its best.

The Colorado Territorial Legislature did not enact mining regulations of its own until February 9, 1866. This act, known as the "fifty-foot law," limited the width of claims in Colorado to twenty-five feet in either direction from the center of the discovered lode or vein. The fifty-foot width might sound narrow, but it only defined the surface boundary, not the mineral underground. The law limited the length of a claim to 200 feet, but mining rules and customs recognized the miner's extralateral right to follow his fissure vein underground wherever it led him. And the fifty-foot law contained an additional, important provision concerning the length of claims. It permitted an association of up to fourteen persons, all calling themselves locators, to claim up to a maximum length of 3,000 feet. This provision was a bit awkward because it opened the door to fraud. When the Consolidated Montgomery District, the closest mining district to Plummer and Myers's discovery on Mount Bross, approved its own bylaws on March 24, 1866, it sidestepped this "fourteen locators" rule by establishing the size of mining claims that one locator could file as 50 feet by 2,000 feet for each location.[6]

Further regulation arrived in July 1866, when Congress passed the first federal mining law regulating lode and vein discoveries (underground mining) on the public domain. The law created a method whereby a miner holding a mineral claim on the public domain could obtain absolute title to

Joseph H. Myers's cabin at Montgomery, ca. 1868. Denver Public Library, Western History Collection.

it from the United States (this is known as a "patent"). In addition, the law incorporated the rules and customs preceding it to the extent that they did not conflict with the new law validating both the fifty-foot law and the by-laws of the Consolidated Montgomery District. If a prospector assembled an association of fourteen persons, he could lay claim to 3,000 feet in length.[7]

Plummer and Myers named their claim the Dwight. When they officially filed the Dwight with the county clerk on March 9, 1869, the record listed fourteen locators. The Dwight, 50 feet by 3,000 feet (3.44 acres), descended from the silver outcrop 1,000 feet down the steep northeast face of Mount Bross to the lower end line of the claim.[8]

Immediately following the legal recording of the Dwight, a second legal entry showed eleven of the men named as locators quit claim deeding their interests to the owners of record: Plummer, Myers, and a third man named Richard B. Ware. Everyone knew that eleven of these men were not bona fide discoverers. Ware had either grubstaked Plummer and Myers with supplies or had bought in. One would assume that all these men had to be present to sign the quit claim deed, but the records do not show it. The other locators may have been nothing more to the county clerk than known residents of Fairplay. They may have never even known that they were listed as locators, but the quirk in the law made everyone go through such a charade.

Daniel Plummer, Joseph H. Myers, and Richard B. Ware now controlled a mining claim at 13,600 feet on Mount Bross. At 14,172 feet, Bross is the southern summit of a huge mountain massif. The northern summit, Mount Lincoln, is higher still at 14,286 feet. These two summits are connected by Mount Cameron, which, at 14,238 feet, rises above the ridge crest one mile north of Bross and half a mile west of Lincoln.

The men were not unaware of the hurdles before them and knew that finding silver in an outcrop at 13,600 feet was only the beginning. Plummer and Myers next needed to determine the value of the silver in the outcrop by having a sample of the ore assayed. They decided to mine 100 pounds of ore from the Dwight as their sample and ship it to Newark, New Jersey, for an assay. When they received the results, three assays of the ore sample proved very rich, yielding at a rate equaling 265 ounces in silver to the ton ($481.80 per ton with silver at $1.32 per ounce), 314 ounces ($414.48 per ton), and 400 ounces ($528.00 per ton). This was welcome news because, in Colorado's limited ore market, ore yielding less than 200 ounces in silver per ton was not profitable.[9]

Historical accounts do not make clear why these men shipped the ore all the way to New Jersey, but the Territory of Colorado offered such a limited ore market in 1869 that they may not have had much choice. Transportation challenged every activity in isolated areas in the mountains. At this time the routes into northwestern Park County followed either a 105-mile wagon track from Denver or pack trails over several mountain passes through Summit and Clear Creek Counties. Every necessity of life cost dearly. Ore from mines was heavy and expensive to transport.

Their 100-pound sample had proved very rich, but this was still too small an amount to indicate much. The men faced considerable financial challenges before they could know how much of this rich ore the Dwight contained. To continue mining, they would first have to transport the minimum requirements for human existence—shelter, water, food, and fuel—to this high altitude, and that required great effort and organization. Providing minimal basic necessities to make life possible for men attempting to work in such an unfriendly environment would be easy compared with the enormous physical and psychological challenges of actually mining the silver. And hauling expensive mining machinery, equipment, tools, powder, and more fuel would translate into a monumental undertaking.

In light of all this, Plummer, Myers, and Ware, who were not rich men, shipped no more ore while they contemplated what to do next. The Dwight remained just another undeveloped mining claim, another prospect hole like thousands in Colorado. It was not until early in the summer of 1871 that

Mounts Bross (left) and Lincoln (right). Montgomery Reservoir (foreground) covers the townsite of Montgomery. Photo Harvey Gardiner.

Plummer and Myers climbed back up Mount Bross to the Dwight. They then discovered even more indications of silver in the limestone near the Dwight, but slightly higher at about 13,700 feet. With keen prospectors' eyes, they traced the silver deposit horizontally along Mount Bross. Following the contour of the mountain, they traced the silver to where it intersected and crossed the Dwight in the area where they had found the original outcrop. Remembering how rich the 100-pound sample they had sent to New Jersey proved to be, they became curious about this new silver deposit.

They needed more ore samples and enough of the ore so that they could clearly understand what they had found. But in order to mine the samples,they would have to spend several days in a barren above-timberline environment that offered no shelter. At this altitude, the intense brightness is untempered by shadows or trees, with twice as much ultraviolet radiation and 25 percent more light than at sea level. The thin air holds less heat, and even in early summer powerful thunderstorms with cold, hard-blown rain or snow rake the mountains. What the weather may have thrown at them while they mined their samples is not known, but they were certainly lucky in that they had no competitors on Mount Bross. In fact, Park County mining had sunk to such an ebb that no silver mining activity had occurred since they had shipped the ore from the Dwight in 1869. They once again took ore samples, shipped them, and awaited the assay report. The results from the second samples proved more than just encouraging. By mid-July Daniel Plummer, 49, and Joseph H. Myers, 36, found even their wildest dreams realized.

News of this nameless rich strike quickly spread in this portion of Park County. Those who weren't making preparations to climb Mount Bross to do some prospecting were no doubt thinking about it. Events took on a hurried, excited pace. Not more than two weeks after Plummer, Myers, and Ware received their test results, the first news account appeared in the *Denver Tribune* of July 28, 1871.[10] This article appeared even before the men could officially file their claim, but that caused them no problem because the min-ing law allowed them sixty days after discovery to locate their claim.

On August 5, 1871, they filed the "Moose Silver Ledge or Vein." Using fourteen names as locators, they preempted 50 feet by 3,000 feet (3.44 acres). Drawn in an unusual dogleg manner, the Moose horizontally followed the contours of Mount Bross. Five days later, on August 10, the *Tribune* wrote that six assays of ore from the Dwight and the Moose had ranged from a low of 141 ounces in silver per ton (worth $183.49 per ton, with silver at $1.30 per ounce) to a high of 879 ounces in silver per ton (worth $1,143.49 per ton). This was very rich ore. A flood of excitement hit the area.[11]

Burros hauled ore down from the Moose to the base of Mount Bross. This ad is for an ore packer in Georgetown. *Mining Review,* August 1874.

At this point Plummer, Myers, and Ware suffered the usual financial predicament of prospectors—they lacked the money necessary for supplies and equipment to start mining their claim. But rather than sell out for a sum that might later prove a pittance (as many prospectors would have done), they chose to bring in partners who could provide working capital.

The primary partner they found, Judson H. Dudley (1832–1900), had arrived in Colorado in 1858.[12] One of the organizers of the Denver Townsite Company, he operated a freighting business between the Missouri River and Denver. From 1863 to 1865, when large amounts of mining machinery bound for Colorado came across the plains and Indians disrupted communication between Denver and the East, exorbitant delivery charges ensured that at least the freighters were making a lot of money. Judson H. Dudley was one of them, and he had his eyes open for ways to invest his money. Through Dudley, another man, Andrew W. Gill of New York City, also appeared as a partner. This new partnership became a legal matter of record posthaste when Plummer, Myers, and Ware deeded an undivided one-half interest in the Moose and Dwight to their new partners on August 23, 1871.[13]

At the outset—and before a lot of other prospectors got in the way—the five men secured control of more ground in the area of the Moose and Dwight. On September 2, 1871, they filed three more fifty-foot claims: the Dudley (50 feet by 1,600 feet, 1.84 acres), the Gill (50 feet by 1,800 feet, 2.15 acres), and the Bross (50 feet by 2,000 feet, 2.30 acres). Although they continued to file claims, they still had no understanding of the unprecedented nature of the silver deposit. The Moose followed the outcrop horizontally, and they tried to cover all their bases by staking the Dudley, Gill, and Bross parallel to the Dwight descending the northeast face of Mount Bross. The Moose claim intersected and crossed the upper portion of all these claims, as it did the Dwight. These five men now controlled 13.17 acres high on Mount Bross.[14]

They wasted no time. By the end of September 1871 the Moose claim had been surface-stripped for a distance of 250 feet. High-altitude weather in September permitted outside work, and this simple, inexpensive mining un-covered ore from sixteen to twenty inches wide, which assayed at 400 ounces in silver per ton. Ore this rich and easy to mine is good fortune, but the Moose still had one critical drawback: The mine needed a market where the ore could be transported and sold at a profit.[15]

The only large and metallurgically efficient smelter in the territory was the Boston and Colorado Smelter at Black Hawk. The most direct routes to Black Hawk from the base of Mount Bross required crossing either Hoosier Pass or Handcart Pass, then going over Argentine Pass, which at 13,207 feet was one of the highest passes ever used in Colorado. These were pack trails, not wagon roads. A pack train's normal speed is ten to fifteen miles per day. The transportation expense to haul heavy ore over these trails would have been astronomical. The one route to Black Hawk that could handle heavy ore wagons went to Denver, but from there the ore would have to be shipped back up into the mountains to Black Hawk. In addition to this problem, the Boston and Colorado Smelter worked ores from mines in Clear Creek, Gilpin, and Boulder Counties. It very likely would have been backed up with ore, and when it could have smelt ore from Mount Bross was uncertain.

At that time, no smelters existed in Denver or anywhere else on the plains of Colorado. But Denver still offered one definite transportation advantage: a railroad connection. This opened up another market in the East. Plummer, Myers, Ware, Dudley, and Gill decided to mine thirty tons of ore from the Moose and ten tons from the Dwight—the minimum they needed to mine and ship in order to understand the ore in their claims. They had no choice but to risk the expense of mining, shipping, and smelting this amount of ore. Pack trains of burros began hauling ore down the steep mountain trail from the Moose to the wagon road at the base of Mount Bross. The significance of the burro's unique ability to transport heavy ores down the narrow, unstable trails cannot be overemphasized. Theodore F. Van Wagenen, publisher and editor of *The Mining Review*, paid homage to the often overlooked burro:

> With a resignation to fate as a beast of burden, the full-grown burro, or "jack," has a sturdy body supported by short but wonderfully strong legs. Built to carry loads, the beast has small feet that can pick their way safely along the narrowest and most tortuous of trails and up grades that would discourage a horse. Added to this is a mentality to which nervousness and panic are unknown. If the burro's load projects so far beyond his flanks that it collides with obstructions along the narrow trail, he never loses his head as a horse will, nor backs away in surprise or fright. Instead, the burro will either attempt to overcome the obstruction by steady pressure, or stand right where he is until assistance arrives.[16]

"Jack power," burros hauling 200 pounds each, made two daily trips ,carrying ore three miles from the Moose to the base of Mount Bross. In November 1871 the Moose and Dwight shipped forty tons of ore 105 miles to Denver at a freight of twenty dollars per ton.[17] From there the ore went to New York City and then on to Swansea, Wales, as ship ballast.

If shipping the ore from the Dwight to New Jersey had seemed like a long way, shipping ore across the Atlantic bordered on the unbelievable. This

Judson H. Dudley, *The Trail,* October 1911.

involved freight charges, forwarding commissions at Denver, New York City, and Liverpool, and when finally at Swansea, charges to smelt the ore. But there was a logic of necessity in this, because Swansea, one of the most advanced metallurgical centers in the world, had the technology and the capacity to smelt the ore. Colorado's poorly developed ore market in the early 1870s forced many mines to ship ore to Swansea. Only the richest ores could show a profit after deducting the expenses of the long trip. Mine prospectuses tried hard to play down the expense of shipping ore there, and a number of ore shipments from various mines did not even cover expenses. The lack of an ore market forced this on mine owners, but in time only inexperienced investors were fooled into using Swansea as a smelter.

More than just meeting expenses, however, the Moose demonstrated how rich its ores were. Without any sorting or other preparation, the mine sacked and shipped ore to Swansea. The shipping cost about seventy dollars per ton ($2,800 for forty tons of ore). In the first forty days the Moose, a poorly developed mine at 13,700 feet, produced $21,768 in ore. The first forty-five tons (450 "jack" trips and $900 in shipping to Denver) paid an average of $450 per ton. Constrained by isolation, primitive transportation, and a limited ore market that hindered development, the Moose was the archetypal rich mine. Producing bonanza-grade ore from open pits at the very start, the Moose literally paid from the "grass roots," as miners put it.[18]

Chapter Three

Organizing the Moose Mining Company

Bonanza-grade ore eased the Moose's transition from a rich prospect to a producing mine. The ore was so rich that it could show a profit even with only a few hundred pounds at a time transported by burro. But Plummer, Myers, Ware, Dudley, and Gill had two problems. They did not hold actual title to the Moose or any of the other claims. And they needed to organize a mining company before they could start large-scale production.

To secure actual title, the five men had to master the complexities of land policy and mining legislation. According to the Mining Law of 1866, each discoverer had to meet a number of legal requirements. Every mineral discovery had to be marked with a substantial stake, post, or stone monument showing the name of the discoverer, the name of the lode, and the date of the discovery. The mineral claim could then be officially recorded with the county clerk, but within sixty days the discoverer had to sink a shaft ten feet deep, or excavate an open cut ten feet deep. This work demonstrated the sincerity of the claimant, and the discovery was not legally complete until the work had been performed. The shaft or open cut also had to show mineral. This much of the work had already been done on the Moose, Dwight, Dudley, Gill, and Bross claims.[1]

Once a claim was officially recorded and met all the requirements of the law, its owners held a possessory title second only to the preeminent right of the United States. In every following year the Mining Law of 1866 required that ten dollars in labor or improvements be made for each hundred feet of claim to demonstrate the sincerity of the occupant. For the Moose, Dwight, Dudley, Gill, and Bross claims the annual labor, as it was called, would cost $1,140 per year.

Survey No. 93

Mineral District No. 3

PLAT

OF THE

Judson H. Dudley, Andrew W. Gill, Daniel Plummer, Joseph H. Myers & Richard B. Ware,

UPON THE

MOOSE LODE

Consolidated Montgomery Mining District,   Park County, Colorado.

Surveyed by S. A. Safford Dep't U.S. Surveyor

Containing 3 44/100 Acres.

Scale of 500 Feet to an Inch.

Variation. 15° East.

The Field Notes of Survey of the Judson H. Dudley, Andrew W. Gill, Dan'l Plummer, Jos H Myers & Rich B. Claim, upon the Moose Lode from which this Plat has been made, have been examined and approved, and are on file in this office. And I hereby Certify that the Lode claimed is of Silver and the value of the labor and improvements thereon, exceed one thousand dollars, as shown by the report of the Deputy Surveyor, and the testimony of two witnesses; and I further Certify that this claim is not embraced either wholly or partially within the exterior lines of any other claim, nor does it include any portion of any other claim.

Surveyor General's Office,
Denver, Colorado,
March 12th 1873

U. S. Surveyor General of Colorado.

The official plat of the Moose Lode. Bureau of Land Management, Colorado State Office.

By performing this labor, Plummer, Myers, Ware, Dudley, and Gill would maintain possessory title to their mine claims. But possessory title was not actual title, and with valuable mining claims, no one wanted any legal ambiguities. The United States has never granted actual ownership of public land but instead has required a "fee simple" title. This means that money must change hands, whereupon the purchaser applies for a "patent," which represents the actual title. To gain a patent on their claims, miners had to adhere to the requirements of the Mining Law of 1866. Thus Plummer, Myers, Ware, Dudley, and Gill began the process of getting their patents. From the start, they did not consider their mining claims on Mount Bross as equals. There was the Moose, and there were the other claims. They began the process with the Dwight, Dudley, Gill, and Bross claims. Daniel Plummer,with power of attorney from his partners, applied for mineral surveys, plats, and patents for these four claims on November 11, 1871.

Applying for a patent meant that the claim would be surveyed and the boundaries marked with monuments by a Deputy U.S. Surveyor. This survey, however, could be done only with an order from the Surveyor General's office in Colorado Territory, and only after the patent application had been perfected. In the application process, public notice had to be made of all mineral claims, so that any other party claiming ownership could file an adverse claim to block the issuance of the patent. Public notice had to occur in three ways. First, notice of the intention to patent and a legal description of the location had to be posted on the claim for ninety days; two witnesses had to sign affidavits stating where the notices had been posted. Second, the notice and legal description had to be posted for ninety days in the nearest U.S. Land Office—in this case, Fairplay. And third, the notice of the intent to patent and the legal description also had to be published for ninety days in the newspaper nearest to the claim. No newspaper existed in Park County, and thus the notices for the Dwight, Dudley, Gill, and Bross claims appeared in the *Denver Tribune* from November 22, 1871, until February 21, 1872.

A U.S. Deputy Surveyor surveyed the claims on Mount Bross between March 18 and March 25, 1872. The law required a minimum investment of $500 in actual labor or physical improvements for each mineral claim. The U.S. Deputy Surveyor would note what he considered to be the value of the improvements, meaning the cost of sinking shafts, digging open cuts, and erecting buildings or driving tunnels, at the time of the survey. The claimants had to swear that they had expended $500. They also had to swear that they had full peaceable possessory rights to the premises, and that the claims did not contain any other known mine or lode.

These legal hurdles took time and attention. An adverse claim concerning any of these mineral locations on Mount Bross would be unlikely, however, since these were the first claims in the area. When no adverse claims appeared after ninety days, the register of the U.S. Land Office at Fairplay permitted the claimants to buy the land for five dollars per acre, plus five dollars for each fraction of an acre. On April 30, 1872, Daniel Plummer paid sixty dollars for the claims. The purchase price, fees for filing, surveys, and newspaper publication totaled $262 for the Dwight, Dudley, Gill, and Bross claims. When the land office at Fairplay took his money, Plummer technically received fee simple title to the claims. The patent applications then went to the commissioner of the General Land Office in Washington, D.C. Except for any irregularity in the application, the subsequent issue of the formal patent followed as a mere ministerial act. Once a patent had been issued, the General Land Office had no power to cancel or recall it.

The Moose went through the patent process separately under the care of Judson H. Dudley. The patent notice for the Moose appeared in the *Denver Tribune* from February 21 until May 15, 1872. The official survey was not performed until February 7, 1873.

Simultaneously Judson H. Dudley paid twenty dollars for the Moose's 3.44 acres at the U.S. Land Office on February 28, 1873. The total patent expenses for the Moose amounted to seventy-four dollars, and the application went to Washington, D.C., on March 14, 1873, nearly a year after those for the Dwight, Dudley, Gill, and Bross. It usually took a long time before the official patent document was issued from Washington, D.C., but the Moose patent went through with lightning speed. Displaying a special status, and probably aided by a Washington patent attorney, the Moose patent cleared Washington in only forty-three days. It was mid-June before the four earlier patents arrived. At a total cost of $336, Plummer, Myers, Ware, Dudley, and Gill had patented five mining claims on Mount Bross.

In the midst of the patent application process, the five men incorporated the Moose Mining Company on April 15, 1872. Six trustees oversaw the company. John McNab (1815–1901), president of the National Fulton County Bank of Gloversville, New York, became the new partner.[2] McNab manufactured gloves and mittens, and he was a wealthy man. The new Moose company had a capital stock of $100,000 in 1,000 shares at a pricey $100 each. In a final legal act, the men deeded their interests to the company on April 20, 1873.[3]

Chapter Four

Opportunity Knocks: The Start of a Mining Boom

At the time of the Moose discovery, no extensive mining was under way anywhere in Park County. Placer mining continued to take place, but only seasonally, because it depended on a flow of water to wash the gold from the gravels in the streambeds. According to Park County Pre-Emption Book D, prospectors had recorded only seven lode claims with the county clerk from January to July of 1871. But as word of the Moose spread, the thought of spending a few days high up on Mount Bross prospecting crossed the mind of everyone in the immediate area, and by the last week of July, local prospectors began staking off more discoveries. Opportunity knocked as the mining excitement slowly built into the great wave of a boom, ushering in the saga of opportunity and chance.

Opportunity also presented itself to any outsider adventurous enough to come to the Mount Bross area immediately. One of the first to arrive was George W. Brunk (1839–1910).[1] Originally from New York State, Brunk had come to Central City, where he worked as a miner and teamster. When he heard one of the first reports about the Moose, he quickly brought his wife and three daughters to Park County.

Near the end of July Brunk began prospecting with Assyria Hall near the Moose. Hall (1841–1917)[2] had been around Buckskin Joe since the boom placer-mining days of the early 1860s. He had served as the Park County sheriff in 1870, and his mud-covered cabin with a barn-like front door was a bit of a local curiosity. On August 7, Brunk and Hall filed their first claim along with Joseph H. Myers (of the Moose) on a lode below the Moose, which they named the Alps. Two days later Hall and Brunk went on to claim the Hoosier lode, situated just below the Alps. Brunk then turned

Alps Mine, November 2, 1875. Charles S. Richardson Collection, Denver Public Library, Western History Department.

his attention to the area south of the end line of the Moose, where he discovered the Hiawatha, which he singly located on August 12.[3]

Prospectors who had come to the area soon discovered that silver-bearing limestone outcrops also appeared on Mount Lincoln. The eastern slope of Lincoln, dominated by two spurs, or ridges, divides as the spurs descend eastward from the summit. These spurs are always referred to as the southeast spur and the northeast spur (by their direction of descent). Addison M. Janes laid claim to the first lode on Mount Lincoln with his Wilson lode, which he discovered on the northeast spur on July 23.[4]

As the chronicle of the start of the mining excitement, Park County Pre-Emption Book D does not always show the date of discovery, but it does show the order in which mineral discoveries were recorded following the Moose. Whereas only seven lode claims had been recorded up to July 1871, twenty-one claims were filed in August on Mounts Lincoln and Bross. This pace of new lode discoveries continued in September. On Mount Lincoln the silver discoveries marched up the mountain from the Lincoln lode at approximately 13,200 feet to the Present Help at 14,157 feet. Janes and William E. Musgrove located the Present Help on September 3. That claim included the summit of Mount Lincoln at 14,286 feet, making it the highest mine claim ever filed in Colorado. On Mount Bross, the discoveries circled around the Moose Mine when Daniel Plummer located the Dudley, Gill, and Bross on September 2, and Hall and Brunk located the Silver Star, which partially adjoined the Moose, on September 22.[5]

Even as claims were being filed, many prospectors were more interested in selling their claims than in working them. This made for a speculative market, which attracted men with money. The first of these wealthy men to appear and buy mining property was William H. Stevens (1820–1901)[6] of Detroit. Born in Geneva, New York, Stevens had moved with his parents to a farm in Wisconsin. After the discovery of copper in the Keweenaw Peninsula in 1841 in Michigan, he worked in the Lake Superior District copper mines. In 1847, when the U.S. survey expedition led by Dr. Charles T. Jackson came to the Lake Superior area to survey the copper deposits, Stevens hired on as a woodsman so he could observe the geologists and increase his knowledge of mineralogy. This led him to be a "land looker" and prospector for a group of Boston investors. In time Stevens said he had leased 100,000 acres of mineral land in the Lake Superior District. He went into business for himself and in his first twelve years made a profit of $96,000. By the time he quit the Lake Superior District, he had made $250,000.

A successful "capitalist" and experienced mine scout, Stevens then began evaluating mineral lands in Colorado, Montana, New Mexico, and Utah for potential purchasers. He also looked for good mining investments for himself, and he had earned the luxury of being able to return to his home in Detroit every winter. Stevens made his first visit to Colorado in 1864, inspecting and reporting on mines for men he knew in Philadelphia. He returned to Colorado in 1865 and visited Oro City (present-day Leadville) in Lake County, which was directly opposite Fairplay on the western side of the Mosquito Range.

Stevens was a hard-bitten man who had worked for everything he accumulated. He liked Colorado and returned yearly from 1865 until 1871. Gruff, blunt in manner, and florid of complexion, with red hair streaked with gray, he was fifty-one years old in 1871. He never seemed to notice the lack of amenities in the West, and he was by that time no stranger in Park County. A gold-mining prospect above Montgomery on North Star Mountain had been named for him on July 26, 1871.[7] This suggests that he had been through Park County early in July, before word of the Moose had spread. After leaving Colorado in 1871, Stevens traveled to Utah to survey the excitement over the silver-lead deposits there. But when he heard about the Mount Bross discoveries, he returned to Park County by mid-August, ready to buy mine claims.

Claims seemed to be waiting for Stevens to snatch up at fifty dollars each from prospectors who preferred to sell claims rather than work them. On August 28, for fifty dollars each, Stevens bought the Wilson and Haines on Mount Lincoln, the Hiawatha on Mount Bross, and his namesake "William H. Stevens lode" on North Star Mountain. In September his fifty-dollar purchases continued when he bought the Baker, the claim adjoining the Moose on the east side, and the Alpine, Miller, Monsoon, Plato, Lime, and Lincoln lodes on Mount Lincoln.[8]

Stevens continued to build the mining excitement when he formulated the idea of an association. He needed partners eager to work together for purposes of mining, working, developing, and selling mining properties on Bross and Lincoln. It took little time for five men to join Stevens to form what would be called the Park Pool Association[9]: Nathaniel P. Hill, owner of the Boston and Colorado Smelter at Black Hawk; Hill's smelter associates, Hermann Beeger (chief metallurgist) and Henry R. Wolcott (assistant manager); Joseph A. Thatcher (1838–1918),[10] Central City banker;

William H. Stevens. Photograph Colorado Historical Society.

and William A. Abbe, mayor of Central City. To form the association, Stevens executed and delivered his Mount Bross and Mount Lincoln mining deeds on September 18, 1871, to Joseph A. Thatcher for $500.[11] Thatcher agreed to hold the properties in trust for the Park Pool Association. The association then issued 400 shares of stock at $100 per share. Stevens owned 42 percent of the stock (168 shares valued at $16,800).

Only fifty-three days after the first public announcement of the Moose discovery had appeared in the July 28 *Denver Tribune*, a powerful combine—the Park Pool Association—entered the field. The Moose Mining Company still owned the largest contiguous block of mining claims, but the Park Pool Association controlled more claims scattered across Mount Bross and Mount Lincoln, hedging their bets that some of these would be valuable. By the end of 1871 the association had also purchased the Hoosier, Silver Star, and two-thirds of the Alps. The asking prices for claims were going up, with these last purchases costing the association $1,000.[12]

The importance of William H. Stevens's role in the early moments of this mining boom was evidenced by the honorary title bestowed on him. Such titles abounded in the West, and Stevens received the honorific of "Judge Stevens" when the U.S. Commissioner of Mining Statistics described Park County's new silver discoveries.[13] With the close of November 1871, prospectors had located sixty-three mining claims on Mounts Bross and Lincoln.

Chapter Five
The Boom in 1872

All the silver discoveries on Mount Bross and Mount Lincoln lay 2,000 feet or more above timberline, where the climate is cold and water not easily obtainable. The short summer season offered limited time for prospecting, building roads, hauling ore, or building dwellings. And yet with anticipation mixed with uncertainty, Mount Bross and Mount Lincoln were poised on the brink of a mining boom.[1]

Initially, this new silver-mining area went by the name "Mount Bross District." After the prospecting expanded onto Mount Lincoln, the name changed briefly to the "Mount Lincoln District" before the area became part of the Consolidated Montgomery District, which centered around the old gold-producing town of Montgomery. To the south of Montgomery a rag-tag collection of fifty log cabins and tents making up Quartzville sprang up on a bench at 11,300 feet on the eastern side of Mount Lincoln. A pretty gulch there offered plenty of water and firewood, and Quartzville, 2,400 feet lower than the Moose, became the closest habitation to the high mining operations on either mountain. The Moose Mining Company owned a log boarding house for the miners at Quartzville, and each day the miners had to hike more than two miles up to the mine. They would not reach the Moose earlier than 10 A.M. for the day shift, and the climb had so fatigued them that once there they could do very little effective manual labor.[2]

The amenities at the mine were minimal, and as the mining got under way in 1871, winter also approached. The company wanted to continue working through the winter, but the climb to the mine would become virtually impossible once snow and frigid temperatures set in. A solution was to build an eating and lodging house at the mine. They also started an adit, or tunnel, which offered protection from the elements, and thus the mine could be worked through the winter. All other high-altitude mining on Mounts Bross and Lincoln ceased, and everyone temporarily abandoned Quartzville. In a letter written from Fairplay on December 4, 1871, William F. Baldwin predicted: "The field is open for thousands and next summer will see the sides of the mountains as thickly covered with prospectors as a molasses hogshead with flies."[3]

The winter of 1871–72 proved fierce. Cold temperatures were constant throughout the season, hovering around zero and dropping below zero much of the time. A much larger amount of snow fell on the Mosquito Range than had fallen the previous winter. Reportedly twenty-two feet fell on Mount Lincoln.[4] As spring drew near, no one could be certain whether a rush would bring confusion and trouble in locating new claims. At least one man expected a surge of activity, as he hauled a barrel of whiskey up to Quartzville in March with plans to open a saloon. A *Rocky Mountain News* correspondent was quick to react, noting that "The boundary disputes will be exciting enough, without increasing the flame by alcoholic stimulants."[5]

In a more serious gesture, the surveyor general of the Colorado Territory acknowledged the importance of this new mining area by appointing a territorial assayer at Fairplay in February. Having an assayer nearby would aid mineral exploration by giving the prospectors and miners a description and value for their mineral discoveries. Edward Dyer Peters (1849–1917), a

Edward D. Peters. Photo by Eleanor Bradley Peters,
Edward Dyer Peters (1849-1917) (New York:
Knickerbocker Press, 1918).

graduate of the Freiburg, Europe's most prestigious mining school, arrived from the Caribou Mine in Boulder County to become the first territorial assayer.[6]

An unusually cold and stormy summer followed the winter's heavy snowfall. A letter written at the end of summer summed up the situation: "This summer has been a conglomerate of all that is damnable."[7] But undaunted by the inclement conditions, an army of prospectors, adventurous miners, speculators, merchants, saloon keepers, would-be capitalists, and drifters entered Park County in the summer of 1872, heading for the area of Mounts Bross and Lincoln. They all came with one purpose—to make money. Fairplay prospered and added a bank and newspaper, *The Mount Lincoln Sentinel*.[8] Every day "reports" of important discoveries came down from Bross and Lincoln. "Everybody over in the South Park is crazy, but there is a profitable method in their madness," the *Daily Denver Times* wrote in July.[9]

In spite of all the "discoveries," nothing on either Bross or Lincoln at this time approached what could be called a producing mine, except for the Moose, which kept getting richer. The altitude presented problems, and a

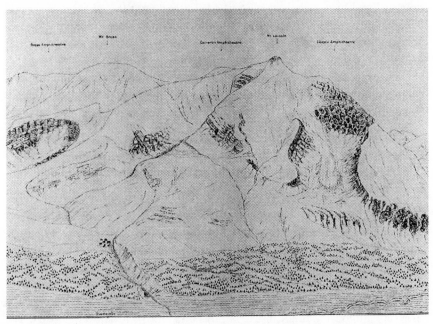

Drawing of Mounts Bross and Lincoln, with Quartzville at bottom left-center. Samuel Franklin Emmons, *Geology and Mining Industry of Leadville, Colorado* (Washington: GPO, 1886).

trail was the only access to the Moose. Fortunately, the surface ore mined easily. By July 1872 they had determined that the ore deposit was four feet thick. Twenty men worked at the Moose, and a week's production could be as much as fifty tons of ore averaging $450 a ton ($22,500 per week). By late summer the surface rock had been stripped horizontally to expose a huge ore deposit fifty feet or longer. Not far below this lay another ore deposit four feet thick.[10] Only the Moose and a few other prospect holes had even been sufficiently developed to indicate what they really were. "When we see a body of ore, we 'gofer' it in every sense of the term," one letter noted.[11] "Depth" had no meaning, because the miners were simply stripping off the ore in open cuts. An open cut requires no underground work until the amount of overburden to be removed makes underground mining more economical. While this surface mining can easily and cheaply remove decomposed ores, it does not develop a mine for the long term. The weather at this altitude limited surface mining to seasonal work. Underground workings, such as the adit at the Moose, were needed to make conditions more hospitable.[12]

Talk of new discoveries swirled around, sounding like rounds of one-upmanship. One rumor, for example, held that eight feet of solid silver had been found in a new and hitherto undeveloped mine on Mount Lincoln. Specimen assays of picked ore yielded as high as 3,000 ounces in silver per ton. But a specimen assay is not a representative sample and is really of no value other than to show the contents of the specimen. With speculation and excitement threatening to throw any sense of reality to the wind, the *Denver Times* compared the area to the prolific Comstock in Nevada:

> People who visit the region where these silver mines are being devel-oped, pronounce them the great silver veins of the continent, and confidently predict that the Emma and Comstock will be compelled to take a back seat in public estimation.[13]

By late July, 400 to 500 miners and prospectors had come into the area. Nine out of ten of these men wanted to discover their own rich silver mines, not work for wages. The scarcity of labor became such a problem by August that Judson H. Dudley had to offer three dollars a day plus board to get employees for the Moose Mine.[14]

Many more discoveries were made in 1872. George W. Brunk and Assyria Hall located the Kansas, Collins, and Dolly Varden very close to the Park Pool Association's Hiawatha on Mount Bross. Instead of offering these claims to the Park Pool Association, however, they wisely held them for themselves. The Dolly Varden would prove the one to make them rich.[15]

The silver mines in the Consolidated Montgomery District produced only a small amount of ore in 1872. Yet for the modest amount of work, this area probably produced more ore than any other silver-mining district in Colorado. Every necessity for living and mining cost more than it should. Men for hire were scarce and the wages offered high ($3.50 to $5.00 per day). By October 1872 the Moose Mine had become decently "housed in" with a boarding house ("Plummer's Boarding House") at the mine and a large shed containing storerooms and a blacksmith shop. No stoping was yet under way in the mine; all the ore came from the drifts, shafts, and cuts that had developed the mine.

Accounts of how much money came out of the Moose are all second-hand, with one stating that by late summer the mine had taken out $75,000 in profit. The Fairplay *Mount Lincoln Sentinel* said that the Moose in 120 days produced 570 tons of ore worth $330,000 ($579 per ton). All the first-class Moose ore still went to Europe, with one shipment assaying at an astronomical $5,200 per ton. The Park Pool Association's Baker and Hiawatha had between them produced 300 tons. As the winter of 1872–73 descended on Mount Bross, the Moose and the adjacent Baker Mine were housed in for the winter. All other mining shut down.[16]

Chapter Six

Silver Replacement Deposits and the Mining Law

T he silver on Mounts Bross and Lincoln occurred in horizontal pockets, chambers, and irregular masses—the likes of which had never been seen in Colorado. Miners in Park County, as elsewhere, expected ore to occur in fissure veins following fractures and faults that plunged almost vertically into the earth. These unfamiliar deposits of silver ore puzzled them. The Mining Law of 1866 had not anticipated such horizontal deposits, either, so this new form of ore deposition opened the door to creative interpretations of the law.

The key to the geology of this new ore deposition lay buried in the stratum of grey porphyry that today still caps the summits of both Bross and Lincoln. The porphyry cap, called "Lincoln porphyry," is best developed on the summit of Lincoln. Immediately beneath the Lincoln porphyry lies a stratum of limestone, which for its prevailing dark blue color is called "blue limestone" or "blue lime." In contrast to the porphyry, which is impermeable, the limestone is brittle and soluble. Underground fissures, fractures, and faults break the limestone into small fragments for considerable distances.[1]

Ore-bearing solutions—literally hot springs heated by magma—rising from deep in the earth over millions of years formed horizontal deposits. As these solutions rose to the level of the Lincoln porphyry, the porphyry forced them to spread laterally into the shattered zones, slowly dissolving and replacing the limestone. In its place the solutions left behind silver ore deposits ranging from the size of a hat to thousands of square feet. Because the silver ore–bearing solutions literally replaced the limestone, these deposits became known as "replacement" deposits.

The blue limestone and Lincoln porphyry meet at what is called the "contact." Geologically, this is one of the most favorable conditions for ore deposition, but prospectors and miners had to learn by observation the significance of this geological contact. They would slowly find that most of the silver ore replacement deposits were at or near this contact. Edward D. Peters helped the baffled prospectors and miners by assaying samples of ore, but no clear geological explanation emerged until the Hayden Survey (officially the U.S. Geological and Geographic Survey of the Territories) came through Park County in July 1873. The Hayden Survey produced a geologic map of Mounts Bross and Lincoln, providing the first illustration of the contact between the blue limestone and the porphyry in Colorado.[2]

Many area prospectors had stronger dreams of wealth than practical knowledge of geology. Even the experienced prospector had his work cut out for him in these complex geological formations. The extremely irregular nature of the silver replacement deposits offered the most remarkable regularity and sharpness at the contact with the porphyry. But as the distance from the contact increased, the dissemination of the ore in the limestone became erratic and thinner, and rock with visual similarities was actually very different. Of 433 assays made between May and December 1872, the richness of the yield was comparatively small. Many prospectors, confused by the geology, paid for assays of ore that did not warrant their effort to mine, much less the money they paid for the assay. In contrast, the Moose Mine's ten assayed samples averaged $663 per ton.[3]

While the prospectors and miners struggled to understand the nature of the silver replacement deposits, the county clerk struggled to record the

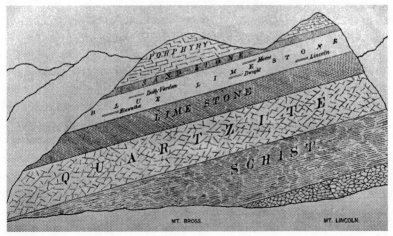

Geological cross-section of Mounts Bross (left) and Lincoln (right). *Mining Review*, September 1873.

location of mining claims. The problem was not only the increasing number of claims being filed but also, in a legal sense, what exactly the replacement deposits in limestone should be called. Normally, mining claims received the label "lode" (meaning several veins found close together). Historically, the terms "lode" and "vein" had been employed interchangeably in both the law and popular usage to describe a fissure vein—a more or less continuous body of mineral lying within a well-defined boundary of other rock. Neither term meant a flat or stratified mass.[4]

Illustrating this confusion is the *Denver Tribune* story announcing the discovery of the Moose, stating that the ore had been uncovered in "chambers."[5] A letter from Fairplay on December 4, 1871, further reveals the confusion surrounding these new ore discoveries:

> I notice that correspondents have described them [the mineral discoveries] as lodes, but I hardly think they are entitled to the name, at least, so far as I could learn, no true fissure veins had been struck, and so doubtful were the claimants of their being lodes, they had recorded them as 'lodes, ledges or deposits.' The latter I believe to be the proper name for them.[6]

Confused about the geological nature of the ore, the prospectors and miners were also uncertain how these silver deposits fit into the law. The Mining Law of 1866 governed lodes and veins (underground mining), and the Mining Law of 1870 governed placer mining (surface mining). These two laws, taken together, divided mineral deposits into two basic divisions: The Law of 1866 applied to minerals "in place," and the Law of 1870 applied to minerals "not in place."[7] The 1870 law applied to the West's early prospectors, with their "wash diggings," who placer mined gold that had been eroded and carried by water from its original source—thus it was "not in place."

Designating mineral deposits as either "in place" or "not in place" left a grey area, as many deposits do not fit easily into these definitions. The distinction between lodes and placer deposits is not clear cut, because it has, in fact, no scientific basis. Nonetheless, the lineal development of a lode held paramount legal importance in the Mining Law of 1866. The distinct individuality of a fissure vein—where the enclosing walls of rock, with the mineral between them, could be continuously traced—established the identity of the lode and, thereby, title to it. This lineal criterion, however, did not work with silver replacement deposits, which by their nature are horizontal and irregular. These deposits had an uncertain dip, strike, or extent, and they had no connection with one another. Unlike fissure veins, these deposits were also impossible to trace. Therefore, silver replacement deposits fell into a legally vague classification of "other forms" of deposit.[8]

Struggling to protect their claims in light of the law's vagueness, prospectors and miners on Bross and Lincoln tried to ensure full legal coverage by using all three terms—"lodes," "ledges," and "deposits"—when they located their discoveries with the county clerk. So it was that the Moose claim had been located on August 5, 1871, as the "Moose Silver Ledge or Vein." Four days later, the attempt at adequate description lengthened with the filing of the "Hoosier Lode, Ledge, or Deposit." Describing these mineral deposits as "ledges," however, did not help, because "ledge" was an Americanism from California that referred to outcrops, and outcrops were assumed to be perpendicular.[9]

To many prospectors, these deposits were "not in place," and they found it logical to start filing their claims as placers. In addition, the Placer Law of 1870 gave a decided advantage to filing as a placer because it established the size of each placer location as a minimum of ten acres in contrast to the Colorado fifty-foot law, which allowed a maximum of 3.44 acres (50 by 3,000 feet). Further, an association of persons could claim up to 160 acres in a placer. A look at two mining properties purchased by William H. Stevens illustrates this "double filing": The Wilson lode was filed as a lode and as a placer claim of ten acres on the northeast spur of Mount Lincoln. The Haines lode also had been filed as a lode and as a placer claim of 40 acres on the southeast spur of Mount Lincoln.[10]

As the mining excitement on Mounts Bross and Lincoln continued to build, Congress enacted, on May 10, 1872, a new mining law to codify, incorporate, and clarify the laws of 1866 and 1870. The object of the Mining Law of 1872 was to dispose of U.S. mineral lands for money value and to declare all lands in the public domain "free and open" to mineral development. This new general mining law permitted a single locator to claim 300 by 1,500 feet (10.33 acres) as his lode, and twenty acres as his placer claim. While Congress did not intend to prescribe law for only one type of lode, the Mining Law of 1872 made no distinction for horizontal ore deposits and continued to use the definition of lodes and veins as true fissure veins. The law also added a new wrinkle to the uncertainty by requiring the prospector or miner to locate his claim at the apex, or top, of the lode.[11]

The apex of the lode may have seemed a logical and fair locator to the members of Congress, but it was irrelevant when dealing with horizontal ore deposits. Since horizontal deposits have no apex, "in place" became that much more important. On Bross and Lincoln, prospectors found much of the ore in a decomposed condition—covered by only a superficial mass of slide rock, debris, detritus, or movable stuff that was easily distinguished from the general mass of the mountain. By this time, the term "placer" had evolved into a generic meaning, which included all forms of deposit other than those occurring in veins. Thus it took the U.S. General Land Office only a brief stretch to construe the mining law to mean that any deposit of any recognized mineral that was not "in place" was intended by the law to be "not in place," that is, a placer. Could the silver replacement deposits therefore be placers?

This may sound like legal nitpicking, but it had the potential to create a legal nightmare when a miner tried to patent silver replacement deposits. All the affidavits required in the patent process were premised on the lineal criterion of fissure veins. A miner could not file truthful affidavits concerning

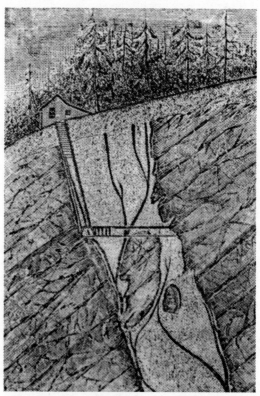

Cross-section of a silver vein.
Engineering and Mining Journal, February 1, 1879.

silver replacement deposits if the affidavits attempted to make these deposits appear to conform to the classification of lodes and veins in the law. If he did this, the miner could be leaving himself open to trouble should the General Land Office in Washington, D.C., later insist upon proper classification, and particularly if title to his property went into litigation. Everybody feared litigation, and confusion in the mining law could prejudice title, thereby depriving an individual of his mining property.

In practice, the Mining Law of 1872 dangled temptation into the confusion over the meaning of "in place" and "not in place." The law allowed the locator of a placer to claim twenty acres; an association could claim up to 160 acres. In contrast, the locator of a lode or vein could only claim 10.33 acres. The ambiguity offered a locator the opportunity to claim considerably more ground and contradicted the fact that it is usually fatal to the validity of mining claims to locate a placer as a lode, or vice versa. Since the mining law did not make it clear what to call these silver deposits, filing as placers offered a chance to gain control of larger pieces of ground. And so it was that William H. Stevens began calling the discoveries placer ground, using the acre method of "ten-acre lots," "one hundred and sixty acres," and "fifteen hundred square feet" to describe the mining claims on Mount Bross and Mount Lincoln.[12]

The confusion encouraged two parties in 1872 to claim, between them, 100 acres south of the Moose Mine on Mount Bross as placers. The first party, Assyria Hall and deputy U.S. surveyor Fred C. Morse (1831–1893),[13] preempted forty acres in October. Their forty-acre Hall and Morse Placer included the Dolly Varden. But since the Dolly Varden had been located as a lode, it could not be relocated as a placer. Therefore, Hall and Morse disclaimed all rights to the Dolly Varden, leaving their placer claim at 38.28 acres. Immediately to the south, the second party, Joseph A. Thatcher and George W. Brunk, along with Henry W. Wolcott of the Boston and Colorado Smelting Company, preempted sixty acres. Their sixty acres included the Hiawatha, which they disclaimed, leaving the Joseph A. Thatcher Placer at 57.68 acres. So it came to be that two placer claims lay a mile and a half above timberline on bleak, dry Mount Bross.[14]

This stretched the definition of "placer" far beyond its intended meaning. But without any specific instructions, the Surveyor General of the Colorado Territory approved the surveys of both placers. And on March 31, 1873, the U.S. Land Office at Fairplay, at the time that the application for patent was made, accepted payment for both. Even though the official patents had not been issued, technically these placers became private property.

The confusion about locating placers continued. In a circular dated July 15, 1873, Willis Drummond, the commissioner of the General Land Office in Washington, D.C., mentioned the "contact-deposits" on Bross and Lincoln. He stated that these forms of deposit could not be given the term "vein," meaning fissure vein, because only excessive excavation could establish whether the deposits were in the form of a true vein.[15] This General Land Office letter did not resolve the confusion about "in place" and "not in place," because it offered no clarification on filing these horizontal claims as placers. William H. Stevens wrote a lengthy letter in reply to Commissioner Drummond's circular, offering suggestions about the classification of mineral lands in the mining laws. Concerning the ore deposits on Mounts Bross and Lincoln, Stevens wrote:

> Owing to the irregularity and uncertainty of their dip, strike or extent and limited product, and having generally no connections one with the other, and not capable of being traced as in mineral veins to establish identity and title,...therefore they were very properly classified with the placer lands and offered for sale in rectangular subdivisions....[16]

Stevens was not involved with the Hall and Morse or Joseph A. Thatcher placers south of the Moose Mine. But with his Michigan friend Alvinus B. Wood and George W. Brunk, Assyria Hall, Fred C. Morse, and three other men, Stevens located, on August 4, 1873, a 160-acre placer claim in the Consolidated Montgomery Mining District on "Buckskin Mountain south spur of Mt. Bross."[17] At this point, between the Hall and Morse Placer, the Joseph A. Thatcher Placer, and now the Stevens/Wood 160-acre placer, the General Land Office in Washington, D.C., woke up to the realization that this confusion had to be settled. In an August 28, 1873 letter, Commissioner Drummond canceled the two placer applications that had been approved. He noted, "It appears that the premises described in this application for patent are found above timberline and some eleven or twelve thousand feet above the level of the sea."[18] Little did he know that the Joseph A. Thatcher Placer included ground at 13,800 feet on Mount Bross.

Hall and Morse appealed Drummond's decision. They described their placer claim as composed of disintegrated limestone and granitic porphyry rock forming a debris from one to fifteen feet in thickness. This deposit and debris were intermixed, showing no uniformity with the ores adrift therein.

> ...deposits are bunches presenting none of the phenomena and characteristics of veins and lodes and that minute particles of minerals are

widely disseminated throughout said rock in place which underlies said drift formation and though the rock in place comes with it small particles of mineral, yet it cannot be said in this formation to occupy any other than a position of bed rock as applied to placer claims.[19]

Hall and Morse argued that Drummond, when he canceled the placer patent application in August 1873, had repealed the Mining Law of 1872 as far as the horizontal silver deposits on Bross and Lincoln were concerned. By their logic, if Drummond maintained that these deposits could not be filed as placers, there was no way to file them. Only one month earlier, Drummond had stated in a letter that horizontal deposits could not be called "veins." In other words, no veins of quartz or veins of other rock carrying silver had been discovered on these mountains. According to the commissioner's logic, they continued, no claim could therefore be entered or patented at all in this area. By his interpretation of the law, the silver deposits were neither "in place" nor "not in place."

This proved a hopeless appeal and was rejected in April 1874. Thatcher and Stevens wisely never even bothered to appeal. The U.S. Land Office in Fairplay refunded the money paid, and after this no one tried to file placer claims on Mount Bross or Mount Lincoln.

While the nature of these silver deposits baffled everyone, the contact between the blue limestone and Lincoln porphyry offered an obvious geological roadmap. Not until the mining activity in Leadville a few years later did it become recognized that if a silver replacement deposit suddenly ended, another deposit might be found by following the contact. The interpretation of the mining law came to be that whatever the miner would follow, expecting to find ore, was his lode. In this case, he followed the contact, which had lineal development and was "in place." By unspoken agreement, and later legal agreement, miners viewed their claims on these horizontal deposits as having vertical lines descending straight down from their end lines and side lines. In this way they owned a piece of surface real estate, and the meanings of "apex" and "extralateral rights" in the Mining Law of 1872 did not apply.

One other creative interpretation of the mining law on Mount Lincoln deserves mention. The Park Pool Association learned an expensive lesson when they contested the necessity of sinking the ten-foot discovery shaft. The association claimed the Eagle lode on the southeast spur of Lincoln by right of discovery in 1872, but they did not sink a ten-foot discovery shaft. They also did not record the location of the Eagle with the county clerk until September 1873. In the meantime, another party developed and officially recorded the Eagle. The dispute went to court, where the association tried to argue that since the silver deposit at the Eagle lay on the surface, it was not necessary to sink the ten-foot discovery shaft. But both mining laws demanded improvements, the initial one being a ten-foot discovery shaft to demonstrate the sincerity of the occupant. The judge removed any doubts about the discovery shaft and the recording of claims when he ruled against the Park Pool Association. The negligence proved costly: The association had to purchase the Eagle lode on July 3, 1874, at a cost of $15,000.[20]

Chapter Seven

Creating an Ore Market

No mining area can hope to sustain itself for long if it depends on high-grade ore. Low-grade ores and base metals form the bulk of all the metalliferous rock in Colorado, and the most solid mining economy always rests on profitably mining large reserves of lower-grade ore. By the early 1870s it was obvious that Colorado needed to establish a mining economy where base metals, particularly lead and copper, could bear a portion of the expenses of mining and smelting. But until this happened, Colorado mining would have to struggle to survive—or better, it would have to survive by mining bonanza ores. Since prohibitive costs handicapped every mineral location except those with bonanza ore, when these ores ran out, mine after mine shut down. It wasn't that they lacked rich mineral, but that they lacked an ore market in which to sell it at a profit.

In 1872 Colorado mining suffered what was labeled "the ore blockade." The primitive animal-based mountain transportation system of trails and wagon roads made the cost of hauling ore exorbitant. The limited ore market wanted only the highest-grade ores. High-grade ore paid the ore buyer and the smelter the largest profit. In this highly selective market, "first-class ores" carrying upwards of 300 ounces of silver per ton found a ready market in Georgetown and Black Hawk. Occasionally, some ore buyers did purchase lower-grade ores in order to get first-class ores, but on the whole, no market existed for ores with much less than 200 ounces in silver per ton. In other words, there was no market for the great majority of the ores in Colorado.[1]

With such a poorly developed ore market, mines were forced to ship ore to smelters outside the territory. The nearest smelters were in Missouri and Illinois, but smelters in St. Louis and Chicago shied away from Colorado silver-lead ores if they contained any zinc. The presence of zinc made it harder to retrieve the lead and silver, and reduced the profit margin. Most of the ore shipped out of Colorado in 1872 went to Europe, most notably to Swansea, Wales, and it was out of the question to send low-grade ores that distance.

To compound the problem, shipping ores out of Colorado drained scarce working capital that could have been invested to help develop a local ore market. In 1872, for example, 2,000 tons of crushed and sacked ore left Colorado. Placing the value of this ore at $250 per ton, these 2,000 tons represented a loss to Colorado of at least $50,000 in working capital.[2] This hemorrhaging of mining capital represents a thumbnail sketch of the problems of all resource-based economies, as described in the Georgetown *Mining Review* in 1873:

> It is undoubtedly a good thing that we have these markets [outside of Colorado] for our ore, if nothing better can be done, but it is proving, as it always has, a losing game, for the country which simply produces the raw materials, and turns it over to outsiders to give it a commercial value, and set it a float in the channels of trade and commerce.[3]

Clearly, Colorado needed to reduce the expense of mining ore, which meant reducing the high cost of everything in mountain camps. It also meant that the territory needed more concentration mills, which would crush and then concentrate the valuable portions of the ore and thereby reduce transportation costs. Colorado also needed more smelting companies to reduce the concentrated ores to bullion and more ore buyers offering a larger, more competitive ore market. In this way, the territory could begin vertically integrating its mining economy from top to bottom by controlling each step from mine to bullion. Until this occurred, however, the limited Colorado ore market sat like a heavy lid atop mining.

The situation began to change as more capital was invested in ore buying and ore reduction. But the transition to a stronger mining economy would take years. In the meantime, many mining companies attempted to get a handle on their expenses and increase their profits by creating their own ore markets. This led to what is known as the small-mountain-smelter phase of Colorado mining.

Illogical as it may seem now, it made more sense to bring the smelter to the ore than to haul the ore to the smelter, and many small smelters were thus built in Colorado's mountains. In fact, by the end of 1872, six smelters had been built in Gilpin County, fifteen in Clear Creek County, two in Boulder County, and two in Summit County. One would also be built in Park County. With the exception of the large Boston and Colorado Smelter at Black Hawk, in Gilpin County these small smelters often had a daily capacity of ten tons of ore or less.[4]

Although small smelters emerged as metallurgical pioneers in the development of the Colorado mining economy, none proved successful for very long. They faced numerous difficulties: What quality fuel was available? How much did it cost to haul fuel to the smelter? Heat alone in the smelter's furnace would not produce bullion. The "charge," the mix of ores and fluxes, had to be correct before the smelter could operate efficiently and profitably. If fluxes were needed, where would they come from, and how much would they cost? Was the smelting process metallurgically correct for the ores being smelted? Were competent management and labor available? Was there money for repairs? Fire bricks, for example, routinely burned out. Was there enough ore available from the mines in the area to keep the smelter running continuously? These were considerable hurdles to overcome, but at a time when Colorado mining faced an uncertain ore market, these smelters often offered the *only* ore market.[5]

Such was the reality facing the Moose Mining Company in early 1872 when it began to consider building a smelter in Park County. No local market at all existed for ores in the Consolidated Montgomery District until half the 1872 mining season had gone by, and the "market" consisted of a small seasonal crushing and sampling works at Quartzville. The Moose could start counting as profit the money the company had been spending to ship ore to Europe should it erect a smelter.

Wasting no time, Judson H. Dudley offered Edward D. Peters, the territorial assayer at Fairplay, the position of metallurgist for the Moose Mining Company. Peters accepted and promptly took on the responsibility for the smelter's construction, which began in July 1872.[6]

Dudley, Andrew W. Gill, and John McNab incorporated the Mount Lincoln Smelting Works Company on October 31, 1872. These men were the money behind the Moose Mining Company, and their Mount Lincoln Smelting Works Company, unconnected to the Moose Mining Company, incorporated to erect a smelter, to purchase and sell ores, and to reduce and treat ores, as well as to buy, sell, and work mining properties. The smelter, commonly called the Dudley Works, fired up on December 1. It comprised a main building (50 by 70 feet), a custom ore building (30 by 50 feet), a blast furnace building (28 by 40 feet), a crushing building (25 by 30 feet), an engine and blower building (25 by 40 feet), a cupel furnace building (20 by 30 feet), a blacksmith shop, a boarding house, and barracks for the workmen. The smelter employed twelve men.[7]

The Dudley Works was a lead-based blast furnace with a daily capacity of eight tons. It needed to smelt a mixture, or "charge," of ores containing a substantial proportion of galena (lead). The furnace's heat would free the lead in the ore, and lead's "self-fluxing" qualities would collect the silver and copper present into the form of a "matte." However, the final parting of this copper matte into bullion could not be done anywhere in Colorado at this time. It still had to be shipped to Swansea, Wales, but the Moose ore contained 7 percent copper, so the copper would pay a great portion of the transportation and expenses.

The Mount Lincoln Smelting Works Company set about securing sources of galena-rich ores for the smelter. On December 2, the company signed three bonds for deed agreeing to buy sixteen galena-rich lodes. All of these lodes were in the Horseshoe Mining District west of Fairplay, and about twenty miles from Dudley.[8]

The Dudley Works had hundreds of tons of Moose ore on hand, enough ore to keep the smelter working through the winter of 1872–73. In a letter to his family, Peters described his surroundings at the smelter on February 1, 1873:

The Dudley Smelter, November 8, 1875. Charles S. Richardson Collection, Notebook 25, Denver Public Library, Western History Department.

It is a beautiful winter's morning and I can hardly realize that I am in the centre of the Rocky Mountains almost eleven thousand feet above you. My office is in a little flat, surrounded by thick pine forests, and just at the very foot of the highest peak in this whole range from British America to Mexico. The ground commences rising from my very doorstep, and in the course of one eighth of a mile, becomes so steep, that only the little Mexican jacks can be used to pack up provisions and supplies to our mines, which are situated almost on the summit of the highest peaks.[9]

He also gave a sense of how critical the smelter was in the ore market, when he commented on his new status:

When I came here I was simply 'Professor' or 'Captain' (every man has a title out here you know). When the furnace started up successfully I was at once dubbed 'Major' and since then have usually been addressed as 'Colonel.' I expect a Brevet Generalship shortly, as promotion is rapid in this country, but if I should ever make a failure it would not take long to degrade me to the ranks.

Another mining company, the Park Pool Association, with three of its stockholders directly connected to the Boston and Colorado Smelter, was also interested in building a smelter, and in the ore market that it would create. Nathaniel Hill sent assistant manager Henry R. Wolcott of the Boston and Colorado Smelting Company to Park County in 1872. After carefully examining the area and evaluating what could become a lucrative ore market, Wolcott recommended that the Boston and Colorado Smelting Company build a smelter in Park County in addition to its smelter in Black Hawk in Gilpin County. The company began erecting its smelter, a reverberatory furnace unlike the Dudley blast furnace, in the spring of 1873 at the confluence of Mosquito Creek and the South Platte River, one mile below the Dudley Works. As of February 1, 1873, only three dwellings stood in this area; but by late May, forty-five houses had been finished or were under construction. The Boston and Colorado Smelter gave birth to the town of Alma.[10]

The prospects for mining all over Colorado seemed more certain in 1873. Companies that intended to buy and reduce ores were beginning to establish a viable market. Only a year after Daniel Plummer and Joseph H. Myers had discovered the Moose, the prospect of two smelters offered the Consolidated Montgomery District a competitive ore market. In the first six months of 1873, the smelters at Alma and Dudley purchased $222,000 in ore.[11]

Chapter Eight

1873: A Good Year Ends on a Bad Note

Very little snow fell on the Mosquito Range in the winter of 1872–73. The "delightful" weather encouraged a building boom in Fairplay that obstructed the main street with building materials in March 1873. The construction included a large hotel or restaurant, a concert hall, a mercantile house with a hall above it, a single-story mercantile house, a large livery stable, and twenty residences. Fairplay now had more than 200 residences, all occupied. The Dudley Works had operated through the winter, and a town, Dudley, had sprung up near the smelter. Dudley had two hotels, one store, two saloons, and a private house, which contained twenty-five bachelors (reportedly all of a good marriageable age).[1]

Starting in March, the emigration into Park County exceeded anyone's anticipation. The stagecoaches on the Denver and South Park Line were so crowded that the company purchased six-horse coaches with a capacity of sixteen passengers. Many men who lacked the stage fare arrived on foot. The *Rocky Mountain News* boosted the prospects of Mounts Lincoln and Bross, saying:

> This district presents more inducements to men with small capital than any other section of the Territory. The ore is found in a lime formation, is very rich near the surface in many of the discoveries already made, and a small outlay demonstrates whether discovery is valuable or not.[2]

Several small encampments of newly arrived immigrants popped up just outside of Alma and Dudley. Miners in Gilpin, Clear Creek, and Boulder Counties prepared to head to this mining boom. A letter to the *Rocky*

Present Help mine, seen from the summit of Mount Lincoln, July 9, 1873. William Henry Jackson, U.S. Geological Survey Library.

Mountain News stated: "We want good strong able bodied men—men who can swing a pick, or roll boulders."[3]

The "balmy" weather also encouraged activity at the mines on Mount Bross. During that mild winter, the Moose Mining Company employed an average of eleven men and reportedly struck a thirteen-foot-thick deposit of solid mineral—a two-year supply of ore. The Park Pool Association had leased the Hiawatha Mine to two men. Leasing, or tribute work, meant that the leasers paid the owners a stipulated portion of the value of the ore that they mined during the term of the lease. Over the winter the men mined 300 tons of ore averaging 201 ounces in silver per ton ($78,390). The association employed six men at the Baker Mine below the Moose, who averaged a production of three tons per day.[4]

On the Dolly Varden claim, the work had opened forty feet of ore six feet wide and four to fifteen feet in depth by May 1873. George W. Brunk and Assyria Hall spent $1,300 in labor developing the ore body, and they sold the forty-three tons of ore for $6,481. When the spring weather began breaking up the snow on Mount Bross, ore from all these mines started slowly coming down the mountain, and the Dudley Works made its first shipment of four bricks of silver valued at $5,000. By June, a hundred men were working underground in the Hiawatha, Baker, Dolly Varden, and Moose.[5]

The Hayden Survey joined all this activity in July 1873. Members of that team noted that at a depth of 50 to 200 feet below the Lincoln porphyry lay a stratum of limestone containing silver and lead. The survey also reported that sixty-two men worked in the area of the Moose Mine, many of them constructing buildings. The Park Pool Association had graded the first wagon road up Mount Bross to reach their Hiawatha and Baker Mines, where a number of men were also building dwellings. "There has been great excitement and hundreds of acres of ground have been taken up," the Hayden Survey noted. "The whole surface is specked with prospect holes, and men with picks and shovels are wandering about everywhere."[6] The mining on Mount Lincoln inspired visitors to hike or ride from Quartzville up to the summit of the mountain. Members of the Hayden Survey made their way to the summit, and a *Denver Times* correspondent did the same a month later. He described sixteen mining properties that he passed while following a zigzag trail across the southeast spur and northeast spur on his journey from Quartzville to the summit.[7]

The lowest mine on the mountain, the Lincoln Mine at 13,200 feet, produced an average of two tons per day with "comfortable quarters" for the men, a blacksmith shop, and an ore house all built of stone. Higher up, five

Fairplay, August 31, 1873. Mount Bross is in background at right. Eliza Greatorex, *Summer Etchings in Colorado* (New York: G.P. Putnam's sons, 1873).

mining claims had been filed at over 14,000 feet. A certain Captain Breece and his associates controlled the Montezuma, Corsican, and Young Hopeful, and also leased the Present Help. Even though the Present Help was on the south side of Mount Lincoln, snow prevented the miners from getting there until June. Once they got there, they graded a road (one and a half miles) from the Moose and Baker Mines, around Mount Cameron, to the Present Help. They also built a frame house twenty-five by twelve feet covering the entrance to the mine and another frame house thirty-nine by twenty feet that served as living quarters. The popular usage of the word "mine" often included mere prospect holes, but the Present Help, at 14,157 feet, had a forty-foot inclined shaft in 1873, and it produced ore. This mine distinguished itself by being the highest producing mine ever worked in Colorado.[8]

More mineral discoveries had been made in Park County during 1873. Transportation improved with the wagon road up Bross, and a road was also under construction up Lincoln from Quartzville. The Moose Mine had more than 445 feet of underground workings, throughout which ore had been found. By fall the Moose had produced $200,000 and sent a 4,700-pound piece of solid silver ore to the annual Territorial Fair in Denver. Judson H. Dudley accompanied this huge specimen to the fair, where he was happy to answer questions about the fabulous Moose Mine.[9]

The Moose, the Park Pool Association, and the Dolly Varden were the major producers in the Consolidated Montgomery District. Mount Lincoln showed a lot of activity—with development work under way everywhere—but developing mines and producing mines being two different things, the actual production on Mount Lincoln was spotty. A comment made by the Hayden Survey indicated both the optimism and the pessimism swirling around the mining boom on these mountains:

> In regard to the limestone silver deposits there is great diversity of opinion and not much confidence as to the permanency of the mines. The pay is exceedingly spotted and variable, but the rocks work easy and there is immense profit where there is any yield at all.[10]

The summer had been very dry. A large forest fire started in late June. With timber in great demand, the fire consumed over 250,000 feet of logs, all the shingle timber, and a great portion of the saw timber that one company owned. Another company lost 300 cords of fuel wood that the Dudley Works had contracted for. The heat and dryness that summer ran hand in hand with the pace of the mining activity. Fairplay, which a year before had been nothing more than an insignificant collection of log cabins, had grown into an embryo city. A sixty-acre addition to the town had been laid out and platted. With the population at about 900, seventy new buildings had been erected in six months' time in 1873. Fairplay boasted ten stores, six hotels, four restaurants, a bank, a newspaper, a concert hall, two churches, a shoe shop, four billiard saloons, two liveries, a planing mill, and five real estate and mining agents. Business lots sold for between twelve and fourteen dollars per front foot, and residence lots went for $50 to $150 each. But the dryness caught up with Fairplay on the evening of September 26, 1873, when a defective stove pipe passing through a canvas ceiling and then up through the roof in the Fairplay House caught fire. The town had no fire-fighting apparatus, and the fire spread resistlessly through the dry wood houses and cabins. About fifty buildings, which included virtually all of the businesses and the *Fairplay Sentinel* office, burned. The lateness of the season gave no chance to recover and scarcely time to find shelter before cold weather set in. Unfortunately, 1873 did not end well, and people had to double up, move to Alma, or leave the area entirely.[11]

Chapter Nine

The Problems with the Ore Market

The blue limestone "contact" on Mounts Bross and Lincoln proved incredibly rich. Park County's silver production amounted to 15,094 ounces ($20,000) in 1871, with all of the silver coming from the Moose Mine and representing only 2 percent of the silver mined in Colorado. In 1872, however, production rose a whopping 842 percent, to 142,209 ounces ($188,000). Then, in 1873, it increased another 116 percent, to 307,633 ounces ($399,000). In only two years, Mount Bross and Mount Lincoln took Park County from its status as a negligible silver producer to a producer of 20 percent of the silver mined in Colorado.[1]

Historically, it is impossible to document the activity in the Consolidated Montgomery District, but these production statistics give a quantifying description of the excitement, anticipation, and hopes of a meteoric mining boom. Led by the mines in that district—primarily the Moose—the silver production in Park County had increased 1,938 percent in two years.

The two nearby smelters at Dudley and Alma gave the local mining economy in 1873 a short-lived competitive ore market. The smelters reduced transportation costs, which lessened the miners' risks. Lower costs would encourage the mining of lower-grade ore, in turn increasing the amount of ore offered to the smelters. Both of the smelters announced their intention to do custom work, meaning that they would buy and smelt ores from all sources for a fee, but two uncertainties existed in this new ore market.

One uncertainty concerned the smelters' ownership. Both the Dudley Works and the Boston and Colorado Alma Smelter were connected to mining companies. They would be obligated to smelt company ore first. Ore from other mines would have second priority, and probably be charged a different scale of prices. The second uncertainty concerned the metallurgical efficiency of the smelters and their ability to compete with each other. Smelting converts ore into a fluid state by means of heat and chemicals, then separates the metals according to their specific gravities. It is an art that demands both theoretical understanding and knowledge of practical details. Edward Peters said that he felt for the Dudley Works the same affection that a mother has for a pet child, but metallurgically he had a hard time providing the proper diet for his child to thrive and grow on. The ore from the Moose Mine contained galena (lead), copper pyrite, silver, zinc blende (a metallurgical problem in those days), and limestone "gangue." Gangue is valueless rock that cannot be avoided in mining valuable ore, but the gangue from the Moose contained an overwhelming mass of barite (silicious heavy spar).

The proper amount of galena in the smelter "charge" would combine with the silver present in the Moose ore. A low-melting-point alloy would form, also combining with any gold or copper present. This molten matte would be tapped off at intervals and run into molds. Barite, however, interfered with this process by making the ore charge harder to melt. The smelter thus had to consume additional fuel to smelt the barite.

In addition to the barite problem, it became obvious that the Dudley Works did not have as much galena ore as it needed in the smelting process. The Mount Lincoln Smelting Works Company had paid $16,000 for eleven lodes in the Horseshoe Mining District in June 1873 that were supposed to supply the critical galena. In reality, the area proved to be poorly developed, the expenses high, and the ore production limited and unreliable during the winter months. The Horseshoe District could not ship enough galena to Dudley. Another problem involved the supply of fuel for the smelter. The Dudley blast furnace needed good-quality charcoal at a reasonable price. But the only raw fuel available in Park County—spruce and fir—did not make good charcoal. These problems pushed the cost of smelting a ton of ore in the Dudley Works to $63.61. This put the smelter at a competitive disadvantage while making it unprofitable to mine low-grade ore.[2]

The Boston and Colorado Alma Smelter avoided these problems by using the "Swansea process," which used a reverberatory furnace. Nathaniel Hill had brought the reverberatory furnace to Colorado following his visits to Swansea.[3] The furnace is an oval hearth covered with a low, arching roof. The roof bends the flame from the furnace at a right angle along the curved ceiling. As the flame follows the underside of the arch, it "reverberates" downward upon the mixture of ores, distributing the burning gases evenly over the hearth. The temperature reaches 1,400 degrees centigrade, and the reverberatory furnace runs well on spruce and fir, consuming from ten to

Boston and Colorado Alma Smelter, date unknown. Denver Public Library, Western History Department.

twelve cords of wood daily. This process also avoided the problem of a galena supply, since it did not use galena as a flux.

Despite owning all the galena prospects in the Horseshoe Mining District, the Mount Lincoln Smelting Works had no choice but to refit the Dudley Works with a reverberatory furnace. Edward Peters began the conversion in August 1873, and when completed, the smelting cost declined 45 percent to $29.16 per ton. The Dudley Works produced $121,184 in matte during 1873, but while Peters retrofitted it, the Boston and Colorado Alma Smelter monopolized the local ore production and began smelting ores from the Moose Mine. After the Dudley Works restarted with its reverberatory furnace, it still had to compete head to head with the highly capitalized Boston and Colorado Alma Smelter for ores and fuel. This proved a losing proposition. Edward D. Peters gave up the struggle, closed the Dudley Works, and left Park County in January 1874.[4]

The reverberatory furnace was in great vogue during the nineteenth century. In time, however, technological advancement of lead and copper smelting in the blast furnace left the reverberatory furnace far behind, and Peters became a world authority on the subject. But in 1873 in Park County, where fuel and fluxes were hard to find, the reverberatory furnace worked best.

How much matte the Boston and Colorado Alma Smelter produced during 1873 is not known. The matte had to be sent to Black Hawk for final refining, and there it became intermixed with silver ores from Clear Creek and Boulder Counties. The Boston and Colorado had eliminated the Dudley Works, but one very loud complaint about the Boston and Colorado remained. When the Park Pool Association organized, it had been assumed that the presence of stockholders who were part of the Boston and Colorado Smelting Company would guarantee that the smelter would treat ores from the association's mines fairly. William H. Stevens, who owned 42 percent of the association's stock, felt that the Boston and Colorado was overcharging for smelting and not paying a fair price for the ores, either. Stevens wrote Henry R. Wolcott, manager of the Boston and Colorado Alma Smelter and an owner of shares in the association, that the smelter was to make a thirty- to forty-dollar allowance per ton on ores from the association's mines. Instead, Stevens claimed, the smelter was charging between sixty and seventy dollars and, in some cases, over $120 per ton. Further, he pointed out that while smelting costs cut the profits of the association's stockholders, the smelter's inside controlling group profited[5]

Wolcott replied that he had made a scale of prices 5 percent lower for the Moose Mine ore than what he paid for ores from the Park Pool Association, but he did this to drive the Dudley Works out of business. "The Dudley works could not afford to do as well, the ore from the works is all being

Henry R. Wolcott. From *Colorado: Deluxe Supplement* (Chicago: S.J. Clarke Publishing Co., 1918).

delivered here [to the Boston and Colorado Alma Smelter]," he wrote, "and they are winding up their smelting business."[6] Wolcott challenged Stevens to find a smelter anywhere in the United States that would offer better buying prices.

Actually, Stevens wanted to sell the mines, not mine and smelt ore, as he stated to Wolcott:

> As regards the policy of PPA [Park Pool Association] it is my opinion and always has been, that there is more money for the shareholders of the Pool to prepare their mines for sale and sell them with a good pile of ores on the dump and a good show in the mine [rather than actually working them].[7]

Stevens was particularly mad that Park Pool Association ore, mined and put on the dump to help sell the Hiawatha, Lime, and Wilson lodes, had been sent to the Boston and Colorado Alma Smelter.

These bad feelings rent the relationship between stockholders of the Park Pool Association. Up to this point, Stevens had been the "unpaid manager." Now, in the summer of 1873, Joseph A. Thatcher dispatched James V. Dexter (1836–99) from Central City to manage the association at Alma. The association paid a 70 percent dividend ($28,000) on its stock. The *Denver Times* highlighted the association's success in an article entitled "Does Mining Pay?" For all of Stevens's grousing, his profit in 1873 amounted to $11,760, and although he continued to hold his shares in the association, he turned his focus to other mining interests that proved of great consequence.[8]

Mining did pay, but the exclusive ore market made the condition of mining in Park County rather disheartening in 1874. The Dudley Works remained closed. The Boston and Colorado worked ores from the Moose, the Hiawatha, and the Dolly Varden. The smelter had so much high-grade ore available from these mines that it only sporadically bought other ores. The Consolidated Montgomery District had no other ore market closer than Black Hawk or the Golden smelter, which had just fired up. This limited ore market could not establish a strong mining economy, and a great deal of dissatisfaction arose over the situation.[9]

Park County miners made two attempts in 1874 to bring competition into the local market. The first centered on Charles Holland, a Chicago banker, who incorporated the Chicago and New York Mining and Smelting Company in February 1874. Holland announced his intention to build a smelter with the capacity of twenty tons per day between Alma and Fairplay at the mouth of Pennsylvania Gulch. When he came to Park County for a visit in June, Fairplay had rebuilt enough from the fire to support about 500 people. Alma had a population of about 300, Dudley seventy, and Holland, three miles above Fairplay, about twenty. Holland busied himself with the construction of his smelter, but when he began purchasing ores, it was only at very low prices; rather than stockpile them for his smelter, he shipped the ores to Golden and other points.[10]

The Holland Smelter was, like the Dudley Works, a lead-based blast furnace. A chemist and metallurgist from Chicago named Dr. Schafer came to fire it up, and then he proudly announced that he had achieved a perfect success in smelting galena ores. Schafer assured all the leading citizens of Alma and Fairplay that he had finally solved the metallurgical puzzle. Claims about smelting galena in a lead-based blast furnace would have received no attention were it not for the arrival of a rookie mining correspondent from one of the Denver newspapers, who received a tour of the smelter. The correspondent knew virtually nothing about smelting yet reported that the correct smelting process for treating Colorado ores had finally been perfected. He even thought that the Holland Smelter was the only lead-based blast furnace in use in Colorado.[11]

No one metallurgical process could reduce the diversified class of minerals found in Colorado. And nobody questioned that a blast furnace could smelt galena, either. The Park County problem was mining and transporting enough galena-rich ores to the smelter. Frustrated by the lack of an ore market, the mining community had been led on by yet another outside expert. In response, the *Fairplay Sentinel* railed about Dr. Schafer:

> Is there ever to be an end...to the influx of learned Professors (?) and Doctors (?), who coming from foreign lands cast their eyes pityingly upon the simple Coloradan in his lowly hut, and patting him on the back assure him that he will tell him how to make thousands out of his dump pile, and tens of thousands out of his gouge? And is there never to be an end of those innocent correspondents, who immediately fall down and worship the learned Professor and tell him that the American is a "worm and no man," who doesn't know of the "genuine German process of chlorination" any more than does an illiterate Fejeean, and to whom the blast furnace and its immense capabilities for reducing "any and all Colorado ores" is an unheard of wonder?[12]

Things were not going well at the Holland Smelter, and they never did. The smelter operated for only three days, later selling for $10,000 at a sheriff's sale in September 1875.[13]

The other attempt to establish a competitive market involved the Rocky Mountain Mineral Concentration Company, which had been organized in Denver to buy and smelt ores. This company had an original scheme and expressed an interest in the Alma area. Mine owners could gain an interest in the company by paying for stock with their ores—even as low a grade as fifteen dollars per ton. In exchange for worthless low-grade ore, the miner would get stock in the company. While it was not clear what would happen with the ore, miners and other citizens at meetings in Fairplay and Alma were willing to grasp at any straw to create an ore market. The miners, of course, wanted someone to build a smelter and then pay them what they considered a fair price for their ore. The ore buyers and smelter interests, for their part, first wanted to see the production of ore before they would be convinced that a market existed. The Park County ore market found itself caught in the push-pull of competing economic interests, with both sides waiting for the other to move first.[14]

In 1874 several mines on Bross and Lincoln, with 800 tons of ore worth $100 per ton, waited for a buyer. The Boston and Colorado Alma Smelter only purchased ore during eight of the fourteen months preceding January 1, 1875. The proposed buying schedule made it obvious that the smelter was only interested in the richest ores. It already had its pick of the Moose, Dolly Varden, and Park Pool Association ores. A bit of the frustration with the ore market appeared in a joke about the Boston and Colorado: "In a case being tried before Judge Brazee [at Fairplay], the counsel asked a witness what was the value of the ore, and received for reply: I have no idea what the value was; I sold it at Hill's works [the Boston and Colorado]; I only know what I received for it."[15] The Boston and Colorado produced $452,000 in matte in 1874. The total silver production increased 8 percent to 333,746 ounces ($426,550), but Park County still lacked a competitive ore market.[16]

Chapter Ten

One Fabulous Mine, One Great Mine, and a Struggling Ore Market

In every mining boom, the area containing valuable ore is over-estimated at first. Reports of wealth from the producing mines, along with the always-circulating stories of new discoveries, fuel the excitement. Scores of promising prospects, however, cannot make as prosperous a mining camp as will two large producing mines simply because an undeveloped mining claim (the proverbial "prospect hole") has no material value until capital and labor open it and demonstrate its intrinsic value. But in a mining excitement, an owner can claim that his prospect has value by virtue of its proximity to the producing mines. In the case of Mount Bross and Mount Lincoln, it could also be claimed that a prospect had value by being located on the same geological "contact" where the producing mines found their ore.

By late 1873 the silver boom in the Consolidated Montgomery District had settled down. No longer was it an area where men with limited means could hope to locate a rich prospect, as claims had been filed on all the promising portions of the "contact" on both mountains. A prospect showing high-grade ore needed to be developed rapidly, and the financial risks this entailed required mining companies that could buy and develop it. Two mining companies with this ability began a process of consolidation on Mount Bross in 1873.[1]

The first of the two, the Moose Mining Company, began to purchase claims surrounding the Moose Mine and the company's other five claims on Bross. The company first purchased the Highland lode in 1873 for $125.[2] Shortly thereafter, seven claims owned by the Park Pool Association on the northeast spur of Mount Bross became available because of the strained relationship among association stockholders. William H. Stevens, as noted, had accused those stockholders who also owned stock in the Boston and Colorado Alma Smelter of price gouging on ores produced from the association's mines. This conflict—and the fact that the expected ore production from the Baker lode, which adjoined the Moose, never materialized—spurred trustee Joseph A. Thatcher to sell all association mine claims near the Moose to the Moose Mining Company on March 6, 1874, for $12,000.[3] In all, the Moose Mining Company had gained control of eight additional mining claims in one year.

The company hired a mine engineer named Charles S. Richardson to survey the Moose's geology and ore deposits and to estimate the mine's ore reserves. Richardson arrived at the company's office at Dudley in July 1875. Born in London in 1815, Richardson had been writing on geological, mineral, chemical, and mechanical subjects for more than forty years. Much of his writing appeared in the London *Mining Journal,* for which he was one of the chief American correspondents.[4]

In the survey notebooks of his work for the Moose Mining Company, Richardson occasionally noted weather conditions, creating a fragmentary picture of Alma and the Moose Mine. As Richardson made preparations to go up to the Moose, he noted in his survey notebook that it snowed hard at Dudley on September 20, 1875. Six inches of snow lay on the ground. Two days later, his baggage headed up Mount Bross, and he followed on horseback. The temperature at the mine stood at thirty-eight degrees at 2 P.M. and by sundown had dropped to thirty degrees. Eight inches of snow lay on the ground at the Moose. Richardson noted that he ate a good dinner that evening but, still acclimatizing to the altitude, he did not sleep well his first night at 13,700 feet.[5]

For more than two months Richardson surveyed the mine, calculating the cubic feet of rock that had been removed and estimating the ore reserves. For his eighty-page report, which included thirty-six geological drawings, he earned $750.[6]

Earlier reports about the Moose, written before Richardson arrived, explained that nearly all the underground work had been done in pay ore—ore that could be mined at a profit. Ore deposit followed ore deposit underground. One huge chamber in the Moose, for example, measured 160 feet long by twenty-four feet high and nearly twenty feet wide. It contained an area of about 76,800 cubic feet (or 6,400 tons) of silver and copper ore with little intervening unproductive ground. As of October 1874, the Moose grossed $20,000 a week. This extraordinarily high production continued in 1875, making the Moose the fifth most valuable mine in Colorado. The Moose Mining Company declared a 100 percent dividend on its stock of $100,000.[7]

Richardson's report describing the workings of the Moose made clear why it was such a fabulous mine. The pits and open cuts marking the site of the earliest discoveries possessed a cubic capacity of 42.5 yards (95.63 tons). The value of this ore had been $54,000, which at $564.68 per ton equaled $47.06 for each cubic foot of rock removed. After the mining went underground, the ratio of productive to nonproductive ground in the Moose was three to one, almost unprecedented in any silver mine in Colorado or Nevada.

According to Richardson, the richest silver, lead, and copper deposits lay near faults and dikes underground in the Moose. A large fault running through the center of the mine in a northeasterly direction created a series of vertically inclined fractures about fifty feet wide. Some of these fractures were open crevices or cavities where all the rock had been broken into small fragments. This was an ideal situation for ore solutions to dissolve and replace the fractured blue limestone. Richardson predicted that as long as the fault continued, large bodies of ore would be found both above and below it.

In summary, Richardson calculated that 5,546.5 cubic yards of rock and ore (12,479.6 tons) had been removed. From the now long-gone production records at the mine, he learned that the ore averaged $151.92 per ton. But what the Moose Mining Company wanted to learn was how much ore remained to be mined. Richardson put the ore reserves in the Moose at $624,790 as of January 1, 1876.

The Moose employed seventy-five men in January 1876 and owned sixteen claims on Mount Bross for a total of 74.88 acres. This large piece

Moose Mine, September 23, 1875. Charles S. Richardson Collection, Notebook 25, Denver Public Library, Western History Collection.

of mining property extended 3,700 feet in length from east to west, and irregularly 2,500 feet from north to south. The operations centered on the Moose, where the bonanzas kept appearing. Only five of the other claims had even been fully opened. The Moose had no shafts. Instead, four working tunnels entered Mount Bross. Two passages, or "winzes," connected the two levels in the mine. The airways through the stopes and winzes down to the second level gave the Moose excellent ventilation.[8]

The steepness of the eastern side of Mount Bross simplified the mining. By driving new working tunnels lower down on the mountainside, the miners used gravity to bring the product of the mine to the surface without hoisting ore up shafts, which required expensive machinery and lots of fuel. The company projected driving a deep tunnel fifty feet below the present workings. This tunnel would hopefully cut the ore chambers, and all the product of the mine would then exit the mountain through this tunnel.

In July 1876 the mine had enough ore in sight to keep twenty-five drills (with two men to a drill) at work. Ten men did nothing but haul ore out of the mine. Every ten or twenty days the mine seemed to make a new strike of ore. The Moose was a fabulous mine: While no records of the production exist, and Richardson offered no specific figure, estimates put the production from 1871 to 1876 at more than $750,000.[9]

Just south of the Moose was George W. Brunk and Assyria Hall's Dolly Varden. Unlike the Moose, the Dolly Varden had started as a bootstrap operation. From its discovery in August 1872 until January 1, 1874, the mine produced $34,798. In 1874 the production had been $19,267 as the Dolly Varden slowly developed into a great mine. The Dolly Varden, like the Moose, had more underground workings in productive rather than barren ground. The mine employed only ten men, but it produced ore running from 102 to 600 ounces in silver per ton. By the lowest calculations, the body exposed represented 3,000 tons of ore, and production in 1875 equaled $36,622.[10]

Hall and Brunk hired Charles Richardson to survey the Dolly Varden after his survey of the Moose. Beginning this survey in January 1876, Richardson found the Dolly Varden ores to be the richest in bulk of any other ore found on Mount Bross. While in the Moose Mine the largest and richest courses of ore lay from horizontal to a pitch of thirty-six degrees, in the Dolly Varden the ore descended vertically along the side of a large forty-four-foot-wide porphyry dike. This dike carried with it an inclined body of white limestone. The ore in the Dolly Varden followed this dike down in nearly parallel planes, with the largest deposits of ore found on the southeast where the white limestone merged into the blue limestone.[11]

Richardson calculated that all the excavations in the Dolly Varden until February 20, 1876, totaled 2,819 tons from mineralized productive ground and 1,611 tons of dead work. Out of 4,430 tons excavated, 64 percent had contained ore, but because no market existed in Park County for silver ores of less than 100 ounces per ton, huge mine dumps of low-grade ore surrounded the Dolly Varden. Richardson also expected that very large deposits would be found at greater depth in the Dolly Varden, and, interestingly, he considered the Dolly Varden rather than the Moose to be the prize of the year on Mount Bross. Not one dollar of debt existed against the Dolly Varden. The mining work had been of the most primitive kind. Two hand-drills in summer and one in winter were the extent of the mechanical appliances used. In winter the men worked only during daytime, while in the summer they worked in two shifts of ten hours each.

Brunk and Hall went on to locate more mining claims near the Dolly Varden. In 1877 they bought the Hiawatha, Compromise, Juniata, and Jo Thatcher claims when the Park Pool Association abandoned Mount Bross altogether. They organized the Hall and Brunk Silver Mining Company in June. By the end of 1877 the company owned twelve claims (86.86 acres) surrounding the Dolly Varden.[12]

By 1876 it was clear that there was one fabulous mine and one great mine on Mount Bross. At the 1876 Centennial Exposition in Philadelphia, the Centennial Commission only awarded medals to ten mines in Colorado. Both the Moose and Dolly Varden received awards, and with another mine on Mount Lincoln, the Russia, these three mines accounted for 82 percent of the silver production in Park County in 1876. The Moose produced $250,000, its ore averaging 157 ounces in silver per ton ($182.12 per ton), which moved it from the fifth richest to the fourth richest mine in Colorado. The Dolly Varden produced $75,397 in ore. The Russia added $47,000.[13]

As the Moose, Dolly Varden, and Russia Mines dominated ore production in Park County, the dynamics of the local ore market continued to limp along. These three mines alone supplied the Boston and Colorado Alma Smelter. Why should any other Park County mine produce ore if it was unclear where to sell it, or what the transportation expenses might be? From the perspective of the ore market, would it be prudent to increase the number of ore buyers before the ore needed to support the market had appeared?

The ore market began expanding after August R. Meyer (1851–1905),[14] another graduate of the Freiburg, became the territorial assayer at Fairplay in February 1874. Meyer traveled to St. Louis in April 1874, where he made an agreement to purchase ores for the St. Louis Smelting Company. During the summer of 1875 Meyer erected a crushing and sampling works at Alma. The Dudley Smelter also attempted to make a comeback in September 1875 after refitting with new furnaces and machinery. The daily capacity at the Dudley Smelter increased from ten tons to twelve tons per day, and the Moose Mining Company proposed to send twelve tons daily to Dudley, instead of to the Boston and Colorado. This would put the Dudley Works at full capacity, leaving little possibility that the Dudley smelter could purchase ore from any other mines before spring 1876.[15]

In October 1875 both the Boston and Colorado Alma Smelter and the Dudley Works went into full production. The Alma smelter paid about $3,000 daily for the twenty tons of ore it smelted. The matte produced came to about $7,000 per day. The Dudley Works shipped out its first ten tons of matte in November. With both smelters operating, 1875 set another record for silver production in Park County. According to the *Mining Review*, the county supplied $618,000 in silver to the Boston and Colorado Smelter at Black Hawk, which represented 46 percent of the silver refined there. Charles W. Henderson, a USGS statistician, later calculated that Park County's silver production had increased another 20 percent in 1875 to 412,022 ounces, or $510,907.[16]

The expanding ore market finally exposed the fundamental problem with the county's mining economy. The Boston and Colorado Alma Smelter

August R. Meyer. Photo Kansas City Public Library.

had only 2,000 tons of ore on hand during the winter of 1875–76. If the new Dudley Smelter proved successful and the Moose Mining Company shipped all its ore there, the Boston and Colorado would lack a sufficient supply. The smelter announced that it might have to close the next spring for want of ore.[17]

Close for lack of ore? Overcapacity was the most serious problem a small mountain smelter could face. One estimate stated that the principal mines on Mounts Bross and Lincoln could produce 300 tons per day if worked at capacity, with the lesser-developed mines adding another fifty tons to the daily total. The actual daily ore production from Mount Bross and Mount Lincoln, however, averaged just thirty tons. Only the Moose, Dolly Varden, and Russia had the money needed to outfit for winter work. The Park County ore market rested on these three mines and a summer mining camp.

During 1876 the smelters at Alma and Dudley sent $458,000 in matte to Black Hawk. August Meyer sent an additional $36,000 in silver ore to St. Louis. While the 1876 production declined from that of 1875, Park County still ranked third in silver production, but in six years' time the boom days in the Consolidated Montgomery District had long passed. Two mining companies controlled Mount Bross. In December 1876 the Moose yielded over $30,000 in ore at an expense of $5,200. The workforce at the Moose rose to nearly eighty men in January 1877. Between May 1, 1876, and March 1, 1877, the mine produced over $300,000 in ore with $234,000 of that being profit to the Moose Mining Company. Yet uncertainty was in the air as control of the company changed hands. All of the original Colorado partners—Daniel Plummer, Joseph H. Myers, Richard B. Ware, and Judson H. Dudley—had sold their interests in the Moose.[18] The change in ownership is obvious from two events. First, the Moose Mining Company increased its capital stock to $200,000, and then in August 1877 the company's stock appeared on the New York Mining Stock Exchange. Andrew W. Gill and John McNab remained as partners while the company came under new ownership in New York City. But the Moose Mine and all the mines in Park County, which were not yielding anything near twenty-five tons daily, and a struggling ore market, soon proved small potatoes compared with what was happening on the opposite side of the Mosquito Range in Lake County. All of Park County would be affected as the future of Colorado ascended into a glorious period.[19]

Chapter Eleven

The Discovery of Leadville, and How It Whipsawed Park County

No one could have foreseen the consequences of the 1873 falling-out between William H. Stevens and the other stockholders of the Park Pool Association. Stevens turned his attention to new mining projects outside of Park County, and to various other mining ventures with Sullivan D. Breece. Breece was an old hand from the placer mining boom days in California Gulch, which lies opposite Alma over Mount Evans (13,557 feet) on the western side of the Mosquito Range in Lake County. California Gulch had experienced a gold-mining boom in the early 1860s similar to the booms at Montgomery, Buckskin Joe, and Mosquito in Park County. Placer mining in sections of California Gulch proved fabulously lucrative, yielding upwards of $1,000 per day in gold. As a result, the population surged to 10,000, with the town of Oro City, above later-founded Leadville, servicing the placer mining. The mining booms and large populations at Oro City, Montgomery, Buckskin Joe, and Mosquito explain why Lake County and Park County were both among the original seventeen counties created in 1861 when Colorado became a territory.

Between 1860 and 1864, California Gulch yielded $3,500,000 in gold. But the placer miners encountered two problems: lack of sufficient water for sluicing the sands and gravels, and a dark sand containing iron-stained fragments of heavy rock that filled up the riffles in their sluices. "I was working on claims No. 21 and 22, at that time," Stephen Pease, a successful California Gulch miner, later said, "and the carbonate boulders and gray sand used to cause us no end of trouble. We did not know what to make of it, as it was so heavy."[1] California Gulch quickly passed through the excitement and rush of discovery, attained its maximum production in 1861, and then declined. The early pioneers, content with quickly mining the rich placers, did little exploration on the slopes of the surrounding hills, and the area became one of the many comparatively insignificant gold producers in the West. The large population moved on. By 1874 Lake County miners were working only two lodes and one placer for pay as Oro City faded to a shadow of what it had once been. The dark sand with iron-stained fragments of heavy rock remained nothing more than just that, and the region around California Gulch might have been abandoned soon were it not for the curiosity of two mining men.[2]

Sullivan D. Breece (1815–77), or Captain Breece as he was known, had been a Lake County commissioner in 1861 and a placer miner in California Gulch from the earliest days, owning several claims there. Following the Moose Mine discovery, Breece turned his attention to Park County on the eastern side of the Mosquito Range, where he bought interests in several high-altitude claims on Mounts Lincoln, Cameron, and Bross. His activities in Park County led him to become acquainted with William H. Stevens.[3]

Although active in Park County mining following the discovery of the Moose, Captain Breece never tired of promoting the California Gulch placers. He told Stevens many stories about the boom days there, insisting that because the placers had lacked sufficient water, they had not been worked out. He urged Stevens to take a trip over to California Gulch and make an inspection of the area.

In the summer of 1873 Stevens invited his Michigan friend Alvinus B. Wood (1832–1910),[4] a mining engineer and metallurgist, to Colorado. During this visit Stevens and Wood tried to locate and patent the 160-acre placer on Mount Bross mentioned previously. They also went to California Gulch to look around, and just as Breece had predicted, they concluded that the gulch still had a lot of placer gold in it. That winter (1873–74) Stevens raised $50,000 in the East as placer mining capital. Back in Colorado for the placer mining season of 1874, he and Wood, with Breece as one of the partners, organized the Oro Mining Ditch and Fluming Company. Between relocating abandoned claims and buying claims, including Breece's, the company soon owned all the placers in California Gulch. They solved the water problem by having a twelve-mile ditch dug to bring in a steady supply.[5] Meanwhile, Breece and Stevens continued their mining activity in Park County. Breece leased and bonded the Present Help Mine on September 19, 1872—the highest mine on Mount Lincoln at 14,157 feet. In June 1874, Stevens and Breece became partners when Stevens paid Breece $10,000 for a half interest in three of his claims. Stevens patented two of these claims, the Silver Gem, which lay very close to the summit of Mount Bross, and the Burnside, which included the summit of Mount Cameron.[6]

Their Oro Mining Ditch and Fluming Company proved successful once their new ditch solved the water problem, and Stevens and Wood came to understand the nature of the heavy, dark sand that had caused Stephen Pease "no end of trouble." That understanding came one day in 1874 as Stevens and Wood discovered a large piece of carbonate of lead while overseeing miners working the placer gold in California Gulch. The *Engineering and Mining Journal* years later described the discovery:

> The first actual discovery of carbonate of lead in the gulch, i.e. a knowledge of what it really was, was made in 1874. Mr. Wood one day was seated on a rock, suffering from an attack of the nose-bleed, and, as was his custom at all times, when opportunity offered, was keenly observing the various rocks that lay around him. One in particular, from its peculiar crystallization, attracted his attention, and, picking it up, he discovered it to be carbonate of lead. He carried it to Mr. Stevens, who knew at a glance what it was, but the men [their employees] being close by working and, not wishing to excite their curiosity, Mr. S. threw the "stone" away with a wink to Wood to keep mum![7]

Once on their own, the two men began collecting more samples of this carbonate of lead ore. They had them assayed to determine the silver content. Although they found the first specimens to be far richer in lead than in silver, Stevens, the astute mine scout, had his own suspicion of the ore's significance. This he kept to himself. Even though the assay showed that the samples did not contain large amounts of silver, as time permitted, he and Wood searched for an outcrop of the ore. They found outcroppings at several places and, seeing evidence of what they wanted, located a number of mining claims on what became known as Breece Hill and Rock Hill.

Stevens and Wood had discovered that the silver-bearing geological contact between the blue limestone and Lincoln porphyry on Mount Bross and Mount Lincoln penetrated to the western slope of the Mosquito Range in Lake County. Geologically, tremendous horizontal forces, acting at right angles, had uplifted the north-south Mosquito Range from the east side toward the west. This uplift raised the blue limestone-porphyry contact high on Bross and Lincoln, while it lowered the contact on the western side of the range to around 10,500 feet. Stevens and Wood were the first men to tie together the geology of the two sides of the Mosquito Range: Carbonate of lead occurred at the blue limestone-porphyry contact, the contact contained large amounts of silver, and the contact appeared on the western slope of the Mosquito Range as well as on the eastern slope.

In the spring of 1876 Stevens and Wood announced their carbonate of lead discovery. Yet nothing in particular happened following their announcement, for little mining activity and few people were in the area. The two began to develop and work their new claims, the Rock Mine and the Iron Mine. Assays from the Iron Mine ranged as high as 800 ounces in silver per ton.[8] By comparison, the early assays from the Dwight and Moose mines contained 879 ounces in silver per ton.

William H. Stevens's Iron Mine provided the impetus that led to the development of other mines in the area. At this early moment, Stevens and Wood sent a description of their ore to Edwin Harrison of the St. Louis Smelting and Refining Company. Harrison requested his agent at Alma, August Meyer, to examine the Iron Mine ore. Meyer came to California Gulch in August 1876. He shipped some ore by ox team to Canon City, and then to St. Louis. The transportation cost, twenty-five dollars per ton to get the ore to Canon City, and the smelting cost left almost no profit from the shipment. But the ore contained 60 percent lead, with some silver, making it a great flux in silver smelting. Propositions to buy it arrived from St. Louis and Chicago.[9]

Lake County, as had Park County, suffered a severely limited ore market. High transportation and refining costs made even the Alma market a competitor for the ore. One example of this was the J. D. Dana Mine in Lake County, which paid twenty-seven dollars per ton to ship ore seventy miles, via Trout Creek Pass, to Alma. In a direct line over the mountains, the distance from the J. D. Dana Mine to Alma is about ten miles. The time that Alma served Lake County as an ore market, however, passed like a blink of the eye. August Meyer wasted no time in persuading Edwin Harrison to come on an inspection tour of the new camp. Harrison was so impressed that he established Meyer in a new sampling works there, and the St. Louis Smelting and Refining Company built the Harrison Smelter in the new district, which fired up late in 1877. Before long, daily stagecoaches began running to California Gulch. Reportedly, Stevens refused an offer to lease his Iron Mine for $8,000 per day in February 1878 as attention and hype started to swirl.[10]

The Leadville bonanza defied all previous rules regarding mineral in place, or in a deposit. Mineral was found, not by outcrops, but by sinking shafts down to the level of the blue limestone–porphyry contact that lay like a blanket, or a magnificent carpet, of great thickness. The contact, which appeared in noticeable surface outcrops at high altitude on Mounts Bross and Lincoln, made up only tiny outcrops on the surface in Leadville in comparison to the large area of the contact found below the surface. By

May 1878 the horizontal ore deposit at Leadville had been traced for 3,000 consecutive feet, showing ore from one to six feet deep for the entire distance. This permitted prospectors and miners to locate hundreds of claims on it. Unlike the Comstock in Nevada, for example, where the meaning of "apex" and "extralateral rights" gave legal control of silver in fissure veins to a few mining companies, Leadville was a treasure trove of opportunity. The meaning of "apex" caused hundreds of lawsuits at Leadville, but there was no more an "apex" to the deposits than there is an apex to a floor. Finally, the U.S. Supreme Court ruled that the Mining Law of 1872, as it concerned "apex," was not applicable, and sidelines had to be observed in the Leadville district. This ruling solved the problem that had first appeared on Bross and Lincoln in 1872.[11]

Leadville made "carbonates" a regional term. Carbonates referred to limestone, which is a carbonate rock. It also referred to the lead-silver carbonate ores, to Leadville as the "Carbonate Camp," and to the miners who became wealthy there as "Carbonate Kings." But the contact at Leadville between the porphyry and blue limestone (also known as Leadville limestone or Leadville dolomite) was identical to the contact on Bross and Lincoln, only much more expansive and much better hidden.

Early in 1879 the *Denver Tribune* published a lengthy interview with Stevens titled "Leadville's Columbus: A Talk with Mr. W. H. Stevens, Who Discovered the Great Treasure Vault."[12] Stevens did not mince words about the discovery:

> "People have an idea that the lead carbonates were discovered by pure chance," said Mr. Stevens. "Such is by no means the case. There was no chance whatever about the discovery. I worked as intelligently, and was almost as sure of the result as I now work on my well defined vein, and is now the certain prospect in obtaining mineral. I am not a haphazard miner, but believe in the application of science in prospecting for mineral as well as in its treatment when obtained. If we had more intelligent prospecting and less looking for luck work, the mineral wealth of the State would now be doubled."

Stevens also spoke about understanding the importance of the contact between the limestone and porphyry:

> "It was," said he, "just like it was in Utah, in Montana, in every place I had been from Northern Montana to Mexico. The mineral lies between porphyry and the limestone; it is always found at the

contact. It was the same stuff that I had seen in Utah and in Montana. Why, there are beds of it there which if I were to tell of, would turn all the present excitement from Leadville to that section; but I wouldn't desire to do that, because the mining of it would not pay when railroads are so far distant."

While historical hindsight has the advantage of twenty-twenty vision, it seems there was a particular omission made either during this interview or in the written account of it. In talking about Utah and Montana, Stevens touted his extensive prospecting experience, but if it was obvious that the mineral "lies between porphyry and the limestone," equally obvious was that the closest point to Leadville that this contact had been located and worked was on its eastern limb on Mounts Bross and Lincoln. In other words, Stevens "knew at a glance" what the piece of carbonate of lead Woods handed him was, because he had experienced its outcropping on Bross and Lincoln during his involvement with the Park Pool Association. Stevens had been associated with the mining on these mountains for years, and through him, the discovery of the Moose Mine, only ten miles from California Gulch, had clearly shown the way to Leadville.

Certainly Leadville would have been discovered at some point without Stevens, "the wizard of California Gulch," and his Iron Mine. The most telling compliment to Stevens, however, came in the Leadville *Evening Chronicle* of April 13, 1889, saying that Stevens "read the structural geology of this district with considerable accuracy, even at that time."[13]

Leadville has been written about many times, and it is not the subject here. While Leadville helped mining all across Colorado, it whipsawed Park County economically. The rush to Leadville in 1877–78 depopulated Park County and put its future very much in doubt. Prospectors and miners walked away from their Park County claims. Alma had fifty houses standing vacant by late 1878, with the town little more than a name. Mining production increased in every Colorado mining county in 1878, except for two—Custer and Park. Leadville pulled capital, smelters, and railroads to it like a magnet; mining in Park County all but halted.[14]

The pace of development at Leadville was phenomenal, going from discovery to large-scale production almost overnight. By late 1878, Leadville had five smelters. Capitalists from all over the United States came to Leadville in force, and they invested more liberally than had ever before been known in the history of Colorado mining. This influx of investment capital continued in 1879 when easterners poured in $20,000,000 more. Forgotten in the rain-shadow of this incredible mining boom, Alma and Fairplay had struggled for years, and without much success, to develop a viable mining economy. Now Leadville raked in investment capital, crippling Park County's access to it. Not only did ore production in Park County decline in 1877–78, the ore market structurally worsened.[15]

As the small-mountain-smelter phase of Colorado mining drew to a close, small smelters would still be built in various places, including Park County, at the same time that competition made success even less likely. Eastbound rail freight , for example, had become much cheaper, with Mississippi Valley smelters eagerly seeking ores and paying competitive prices for them. If Colorado smelters hoped to handle the most profitable high-grade ores, they would have to pay nearly as much as the eastern smelters. And in order to be truly efficient, only large smelters working for base metals, in order to get all the values in the ore, could be competitive in the long term. Smelting works needed to locate near cheap fuel and cheap labor, and where ores from many sources could be mixed to flux themselves.[16]

The growth of Colorado mining and the changes in the ore market can be traced from 1872, when the Moose Mine shipped ore to Wales, to 1878, when the Boston and Colorado Smelting Company shut down its Alma smelter and built a new one at Argo, north of Denver. The company determined that it would be more economical to use the coal available in the Denver area than the increasingly less abundant wood around Black Hawk. The company would only sample and buy ore at Alma for shipment to Argo. The Dudley Works also disappeared as a smelter when the Moose Mining Company turned it into an ore concentrator. The Park County ore market now consisted of August R. Meyer and the firm of Berdell and Witherell at Alma, which sampled and bought ore (both were adjuncts of larger operations at Leadville). The Golden Smelting Company also bought ore at Alma. But even when the Boston and Colorado still operated, Park County had no market for ores of less than fifty ounces in silver per ton. No smelter, no transportation system improvements, and additional transportation charges meant that the ore market had distinctly worsened by 1878. After seventeen years, there was still no market for low-grade ore in Park County. Not until 1881, a year after the Denver and South Park division of the Union Pacific Railroad had raced across South Park (with Leadville in its sights), did the transportation system improve.[17]

Chapter Twelve

The Moose Mining Company

T he story of the Moose Mine bifurcates at this point to become the story of a publicly traded New York mining company and the story of a mine on Mount Bross. When the Moose appeared as stock shares on the New York Mining Stock Exchange in August 1877, all four of the Colorado owners had sold out. Reading between the lines, we can speculate that these men may have sold out because they suspected that the known ore bodies in the Moose were becoming exhausted. The remaining partners, Andrew W. Gill and John McNab,[1] had traveled to Colorado several times, but the Colorado partners were the ones who had the mining experience. Now, having no partners with mining experience, the New York–based Moose Mining Company was about to be parlayed into a stock promotion.

Dishonesty and extravagant promises swirled around mining stock exchanges in the 1870s. Brokerage houses, traders, promoters, and insiders who knew nothing about mining manipulated stock prices through newspaper stories about shipments of rich ore and rumors of valuable new strikes. "Experts," who may have been the local saloon keeper or a venerable old-time miner of "forty years' experience," which only meant forty years of failure, made testimonials. Dubious so-called mining engineers also gave their opinions, but the best promotional hook was always talk of impending dividends.[2]

A considerable portion of the eastern money invested in western mining never left the East. Traders beared and bulled stocks with the only extraction being the investors' money. Whenever investors bought mining stock expecting the value of the shares to increase, they moved from investing in mining to gambling in stocks. Some mining companies developed large yields in their mines with a view to profits at the stock board rather that at the mine. And when the dividends paid on stock came from jumps in the stock's value, this had nothing to do with mining.

"It is well understood by most people that 90 per cent of the sales made and reported are bogus," the *Engineering and Mining Journal* wrote about mining stock exchanges in January 1878.[3] Yet ultimate responsibility for fraud rested with investors. Greed for quick and unreasonable profits found the verb "mining" interpreted to mean "large pay on a small investment, with little trouble."[4] Conservative businessmen who would reject highly speculative ventures in other fields found mines irresistible. They plunged into mining: Worthless mining stocks made honest investors unwitting gamblers. As the Latin phrase *mundus vult decipi—ero decipiatur* puts it: "The world wants to be deceived—let it therefore be deceived."

The Moose Mine's appearance on the New York Mining Stock Exchange in August 1877 paralleled this cycle of financial credulity when investing in mining became the sensation in the East. Dull economic times had followed the nationwide financial downturn known as the Panic of 1873. Then Leadville suddenly offered a heady dose of economic optimism, and Colorado mining's reputation for failure from 1863–64 was forgotten. Investors practically stumbled over each other seeking mining schemes. In Boston and New York, mining stock exchanges made a serious effort to interest eastern investors in the stocks of Colorado mining companies. Although the Moose was not a Leadville property, and it predated Leadville, it was, nonetheless, one of Colorado's premier mines.

The Moose Mining Company opened offices at 62 Broadway in New York City in the office of Allen and Stead, a brokerage firm formed in 1875. Daniel B. Allen (1815–1902), president of the Moose Mining Company, had considerable business experience. He was married to Ethelinda Vanderbilt, daughter of Cornelius Vanderbilt, and had been Vanderbilt's principal assistant in the shipping business in the days before Vanderbilt became the richest man in the United States.[5] Allen's partner, Charles M. Stead (1840–1926),[6] was an attorney, banker, broker, vice president of the New York Mining Exchange, and chairman of the Exchange's committee on admissions.

The Moose (stocked at $2,000,000 in 200,000 shares with a par value of ten dollars) opened at $5.50 per share with more than 15,000 shares ($82,500) quickly trading. In January 1878 the Exchange listed sixty-five general mining stocks. Sixteen were Colorado companies, but only eight had paid any dividends. The Moose Mine had paid $175,000 in dividends, more than any other Colorado mine, and represented 31 percent of all the dividends reportedly paid by Colorado mines. The value of Moose stock soon rose to $7.75, becoming the most actively traded general mining stock on the Exchange. When the Moose Mining Company, at the beginning of March 1878, declared a quarterly dividend of $25,000, the stock increased to $8.87. Before March ended, the stock climbed to $9.50 per share.[7]

This meteoric rise to riches in just eight months brought Moose stock under scrutiny. The *Engineering and Mining Journal*, always trying to guard mining's reputation as legitimate business, followed the mining stock markets in each issue. On April 27, 1878, the *Engineering and Mining Journal* declared that the Moose Mining Company's checks would not clear and that its miners and creditors were unpaid. "It is naturally concluded that with such a condition of affairs, the dividend recently declared was not fairly earned," and, the journal warned, "the stock, if not worthless, is certainly an unsafe one for investment."[8]

In short order, Daniel B. Allen, with attorney, confronted the *Engineering and Mining Journal* to declare the statements a lie. David Moffat of the First National Bank in Denver cabled New York that no Moose checks had ever been refused. The journal, however, would not budge, stating that despite "a telegram from one Moffat, Cashier of the First National Bank," the Moose owed the bank between $70,000 and $100,000. This was the money paid as dividends on the stock. The journal went even further, alleging that the First National Bank in Denver had an obvious interest in maintaining the value of Moose stock, since, undoubtedly, it held the stock as collateral for the money loaned to pay the dividend.[9]

One of the most grievous stock crimes for mining companies was overcapitalization, in which the value of the number of shares would far exceed the value of the ore that the mine could ever be expected to produce. The share price of $9.50, for example, put a value of $1,900,000 on the Moose Mine, the same mine that Charles S. Richardson estimated had ore reserves of only $624,790 as of January 1, 1876, and that had produced $250,000 during 1876. "In our efforts to obtain reliable information, we have not had a value of more than $300,000.00 placed on the mine," the *Engineering and Mining Journal* wrote in May 1878, "and in giving it this figure it is placed at a value that is exceeded by but few mines in Colorado."[10] The ongoing discussion of overcapitalization and actual dividends paid led to an immediate decline in Moose stock of 74 percent to $2.50 a share by June 1878.

Meanwhile, at the mine itself, Charles L. Hill still superintended the Moose, employing forty men through the winter of 1877–78. Hill kept a portion of his crew doing systematic development work underground—

prospecting, locating, and blocking out new ore while the remainder of the crew mined ore. In contrast to Hill's approach, many mining companies extracted ore with no thought of prospecting for more. When the ore in sight ended, often no working capital remained for exploration dead work, and operations ceased. Hill oversaw the Moose Mine as a large, well-developed, and well-run mine. From May to November 1878, the best season of the year for hauling ore off Mount Bross, the Moose expected to increase its workforce to eighty men.[11]

Unfortunately, Leadville captured Hill in the summer of 1878. At this point, eastern mining companies often hired and sent out people with no mining experience to manage the mine—a cousin of the company president, for example. But when the Moose Mining Company hired Jacob Houghton (1827–1903)[12] of Detroit to replace Hill, it demonstrated the power and reputation of the company and the mine. Houghton was the youngest brother of Douglass Houghton, the first state geologist in Michigan, and one of the discoverers of the immense copper deposits on the Keweenaw Peninsula of Lake Superior in 1841.

Back in New York City, the Moose Mining Company, in September 1878, changed the dividends reported as paid on the mine to $550,000, a substantial increase from the $175,000 paid up to January 1878. Was this the figure

Charles Hill. From *History of the Arkansas Valley*
(Chicago: O.L. Baskin & Co., 1881).

for dividends paid since the company's incorporation? Wherever this figure came from, it promoted stock sales. The stock rose in six months from $2.50 to $4.70 a share by December 1878. However, the *Engineering and Mining Journal* in its weekly commentary on March 22, 1879, did not let up on the Moose: "Moose continues to hold a very high position for a mine that has done so little, and which does not promise much."[13]

The stockholders were anxious for the dividends to resume, and they forced the Moose Mining Company to issue a statement, which appeared in the *New York Daily Graphic* of February 21, 1879.[14] A promotional piece intended to soothe stockholder feelings, it stated that a primitive type of diamond drill had been purchased for exploration work. But ore production, not exploration work, paid dividends.

Activity at the Moose continued. Since early December 1878, three men had operated this diamond drill from underground setups, and later it operated from the surface high up on Mount Bross. The company had also negotiated a favorable contract with the Boston and Colorado Smelting Company, which caused the closure of the inefficient and costly Dudley amalgamating works. The Moose, at this time, produced three tons of ore daily, which averaged 100 to 200 ounces in silver per ton. Speculation held that if the mine could provide proper accommodations, 150 men would soon be employed, a number almost double the eighty-four men working there in early 1879.

In April 1879 the men at the Moose began to shovel snow off the wagon road, which opened to wagons in May. The diamond drill continued exploration work, and for July the entire product of the mine ran 262 ounces in silver per ton. In August the Moose grossed $40,000. The payroll increased to 115 men in early September.[15]

During the summer of 1879, Leadville egged on wild speculation in mining stocks. New mining companies by the hundred organized, each striving to outdo its rivals in the number of stock shares offered, the number of claims held, and the brilliant show of prospects. It was becoming obvious that the best "mine," for those gifted to work it, was investor confidence— that "mine" cost the least to operate and was practically inexhaustible.

During the week of November 15–21, 1879, the aggregate sales of mining stocks reached a level never before known. Transactions amounted to $40,000,000 a day, and not one-fourth of these mining companies had ever paid any dividends. Moose stock fluctuated wildly between $3.20 and $4.20 a share, becoming the third most actively traded dividend-paying mine on the New York Mining Stock Exchange.[16] When 1879 ended, the

Engineering and Mining Journal, in its yearly review of the mining stock markets, said this about the Moose:

> This stock reached $9.50 in the spring of 1878, under forced dividends, since when there has not been one paid, although promises and flattering reports have been made. For a stock that gives no returns to the holders and really gives no indication of ever giving any, a remarkably high price was maintained during the year. ... Better results are promised in 1880. It would be strange if nearly two years' work did not permit the company to put the mine in such shape as to permit a resumption of dividends after which to look out for an unloading on the part of the insiders.[17]

By the third week of January 1880, transactions in Moose stock became exceptionally large. The volume of these transactions—248,495 shares— was larger than the number of shares owned by the general public. For three days, huge numbers of shares were thrown on the market. Obviously, something was afoot as the price fell to two dollars a share.[18]

Moose stock had become a speculative football for the Exchange to kick about. The price gradually weakened to $1.50 a share, and the Moose

The crew of the Moose Mine, ca. 1877. Mazzula Collection, Amon Carter Museum.

Mining Company deserved the bad press that it received. The press put forward the lame idea that Allen and Stead were forcing down the stock price to "freeze out" certain large stockholders. But the "freeze out" was a blind for a "got out" by Allen and Stead. Daniel B. Allen sold all his stock, complaining that when he had gained control of the Moose Mining Company it had no working capital, and that the mine was in poor shape. The company had then employed a large force of men doing exploration work for a number of months, apparently without success. Allen believed that the Moose was worked out.[19]

Allen may have been the businessman lured by the gamble and glamour of mining, but during his management of the Moose Mining Company, it was almost impossible for the stockholders to get a report on the mine's condition. Frustrated, the stockholders called a meeting for February 24, 1880, in New York City to coincide with Jacob Houghton's presence in town. A committee of the stockholders requested the presence of Allen, Stead, and Houghton at the meeting, but was told to expect a statement in a few days instead. The stockholders did hold their meeting, at which they passed a resolution to have all the stock certificates transferred to their own names. If the need to transfer the certificates to the stockholders' own names sounds peculiar, it shows how slippery the Exchange could be. Allen and Stead

Jacob Houghton. Photo Bentley Historical
Library, University of Michigan.

actually controlled the votes of the stockholders against the real owners of the stock, because most of the stock had been registered in the names of their clerks or brokers.[20]

By March, Moose stock had sunk to $1.10. The once-great Moose no longer presented investing in Park County mining in a favorable light, and now even the *Fairplay Flume* joined the attack:

> Not only has this magnificent property been a stumbling block in the way of the many other mines of the county, by reason of its lack-dividend-paying qualities, but the mysterious way in which all of the applications for information on the part of the stockholders have been met, and the indefinite reports issued semi-occasionally have caused a distrust in the entire district that only the very marked success which attended prospecting last season would have overcome.[21]

Finally, the requested information appeared from the board of directors of the Moose Mining Company. From May 31, 1879, to January 31, 1880, the Moose had produced 1,092 tons of ore, which sold for $102,192.65 ($93.58 per ton). The mine could not begin to meet its promised 10 percent dividends ($200,000 per year). On top of this, the report stated that the ore in one of the largest stopes had become too low-grade to be mined profitably. As of January 31, 1880, the Moose Mining Company was $22,228.66 in debt.[22]

Part of this report summarized a report Jacob Houghton made to the board of directors, in which he expressed optimism that the last fault on the eastern slope of Mount Bross had been encountered. Beyond the last fault, he anticipated a large stretch of ground, sixty acres atop Bross, that would prove the most valuable portion of the Moose estate. At Houghton's direction, the company had located the Mary and Dora claims in 1878 near the summit. The Mary and Dora, along with the Addie claim, located in the area in 1876, became the last properties added to the Moose estate. Without question, Houghton believed in the potential of Mount Bross. He patented several claims—the Crown, Emerald, Reliance, and Argos northwest of the Moose—for himself, and he now proposed that the company explore the area on top of Mount Bross with tunnels driven from the Highland and Julia claims located above the Moose. He concluded his report with a reference to the sixty acres at the summit: "The entire surface is covered with a heavy deposit of porphyry (rhyolite) under-laid with limestone (dolomite) being the same geological formation as those at Leadville."[23]

Ideally, a mine can fall back on rich deposits left undisturbed underground whenever the general average of the yield threatens to fall below a dividend-paying point. These rich spots are called the "eyes" of the mine. But when a mine like the Moose was overcapitalized, pressed to pay dividends, and counted a large number of men on the payroll, the miners picked out the eyes as soon as they found them. Gutting the levels became the order of the day at the Moose. The ore shipments produced cash flow for the company, but only for as long as the ore bodies continued. By February 1880, despite Houghton's optimism, the Moose reportedly had never looked worse. Ore shipments had been virtually suspended. The Moose Mining Company needed to finance exploration work to develop new ore bodies before production could resume. But the company was in debt. No ore bodies stood ready for production. No ore bodies were held in reserve. And the company lacked the working capital necessary to finance exploration for new ore bodies.[24]

When Houghton visited New York in February 1880, he told the board of directors that he needed $25,000 in the next four months for development work in the Moose. In a highly suspicious corporate decision—and although Houghton had requested only $25,000—the board of directors deemed that if $50,000 would not carry the company through, it would create a mortgage of $100,000 by issuing bonds payable in five years at 10 percent per year. The Moose estate was collateral. The stockholders never approved this mortgage, but they really had no choice, either.

When he returned to Park County, Jacob Houghton said that the working capital for the Moose had been raised. In a classic example of the human tendency to lose sight of the real problem by focusing on irrelevant issues, Houghton said that the board of directors wanted to build an aerial tramway from the Moose to Dudley. A tramway would be the most efficient way to transport ore, but he seemed to overlook the reality that first the Moose *needed* ore.[25]

Concurrently, the frenzied speculation in mining stocks reached unnerving levels. Select Leadville stocks, in particular the Chrysolite Mine, commanded forty dollars per share. In the first quarter of 1880, it was estimated that the stock of all the organized Colorado mining companies totaled about $300,000,000. Such overcapitalization meant that paying a 10 percent net return would require an annual production of $30,000,000. To reach this, the 1880 production in Colorado would have to double that of 1879.[26]

This overcapitalization of mining companies and the extravagant promises of dividends meant that mines were under pressure to produce large amounts of ore. The problems at the Moose reflect those of the mining industry as a whole, because no mine could ever continue to produce the volume of ore needed to pay the large dividends promised to investors. At some point, the known ore bodies would be mined out, as with the Moose, and until exploration work located more ore, there would be no stream of money to meet the financial obligations. Two of the biggest mines at Leadville, first the Little Pittsburgh and later the Chrysolite, stumbled in the rush to maximize ore production, collapsing the mining stock boom. On April 17, 1880, the *Engineering and Mining Journal* warned: "For several weeks past, there has been a perfect panic in mining stocks, and many think that the public will cease to interest itself hereafter in mining." Moose stock sank to a low of thirty cents a share on September 21, 1880, with the Moose Mining Company defaulting on its $100,000 bond.[27]

One of the most difficult decisions in mining is deciding when to quit. While ultimately everything depended on the Moose's production, the company in New York and the mine in Colorado seemed almost unconnected to each other. So much money had been invested in Moose stock that a new financial scheme between the Moose Mining Company and the newly created Moose Silver Mining Company of George B. Satterlee and Francis H. Weeks hatched in New York on December 31, 1880.

George B. Satterlee (1834–1903)[28] had been president of the New York Mining Stock Exchange in 1877–79. Wealthy and influential, he headed the New York banking firm of Satterlee and Company. Francis H. Weeks (b. 1844) was an attorney and broker. Their agreement comprises four pages

John M. Fritz. Photo Erik Swanson.

of handwritten legalese in the Park County records. Fortunately, Horace S. Bradford, a broker and a stockholder in the Moose, wrote a letter to the New York *Mining Record* explaining the meaning of the agreement. Bradford said that the defunct Moose Mining Company had no funds in its treasury, and if the Moose estate went to public auction, the stock would be rendered valueless. To prevent this, the Moose Silver Mining Company had been formed to pay the interest on the bonds and to pay off the Moose's floating debt. The Moose Silver Mining Company would also furnish money to work the mine. Within one year's time the new company would give the shareholders of the old company the option to exchange, one-for-one, their shares and bonds in the old company for those of the new. The silver mining company had $300,000 in stock at a dollar per share, and $40,000 in new bonds would be issued in the name of the old Moose company for working capital. If the old company could not repay the bonds, the new company would foreclose on the property.[29]

In January 1881 both the old and new Moose stock appeared on the Exchange. A few shares of the new stock changed hands at $2.40, with the old stock, in sympathy, briefly jumping to $1.90. This whole affair was intended to help inside parties unload their stock. Before long, sales of the new Moose stock became almost nonexistent, having merged with the old stock, which resumed its unpredictable gyrations. During two weeks in October 1881, 205,400 shares changed hands. The stock dropped to as low as sixty-three cents but by April 1882 was selling at $1.70, or 70 percent above the par value of the new stock. No sane investor could possibly consider buying Moose stock.[30]

The *Fairplay Flume* had made a telling comment about the Moose early in October 1880: "We are sure that Mr. Houghton, as superintendent, had administered affairs judiciously, but he could not find ore where it does not exist—and so far no one has been able to say where it does exist."[31] By late 1880 the Moose had resorted to leasing sections of the mine. The company let independent miners scavenge out whatever pay ore they could find. During 1881 to 1882 as many as thirty-six men worked at the mine. John M. Fritz replaced Jacob Houghton as superintendent in 1881, and he did ship several carloads of ore to the La Plata Smelter at Leadville. The total value of the ore hardly exceeded $8,000, however, and Fritz said that while the shipments to Leadville paid all expenses, they left no profits. A strong, steep fault had completely terminated the ore bodies and the ore-bearing horizon in the Moose. Twenty men worked in the mine in January 1882, but diamond drill exploration could find nothing of value beyond the plane of the fault. The truth, which had never been admitted in nearly four years, was that no new paying body of ore had been discovered since 1878.[32]

After years of manipulation of Moose stock, the beginning of the end came when a brokerage firm on the Exchange attempted to raise $40,000 to corner the Moose stock. If it made a successful corner, it could ratchet up the value of Moose shares and, at some strategic moment, unload its shares. The firm failed to raise the $40,000, however, and had to induce its fellow Exchange members to sell short futures contracts for Moose shares, with a pledge to furnish the stock certificates in case of sales.

The prices asked for Moose stock started rising amid this chicanery. Members of the Exchange made sales, but when they called for the stock certificates, they discovered that the brokerage firm behind the manipulation did not possess them. Realizing that they were being duped, members forced the sale of the futures contracts for Moose stock under the Exchange's rules. Arnold and Company, the brokerage behind the manipulation, bankrupted.[33]

The collapse of the "fungus" Moose Silver Mining Company on May 11, 1882, appalled the New York Mining Stock Exchange—or so the press said, at least. Generally, no one gets badly hurt in attempted "corners" except the conspirators themselves, but it took the Exchange a week of unsettled and irregular trading to overcome the Moose collapse. Limping to an inevitable end, the Moose Silver Mining Company in the summer of 1882 sold twenty-two head of jacks (burros), one bay mare, and one bay horse for $730. The mine again defaulted on the interest payments on the $100,000 bond. In November 1882 the *Mining Record* wrote: "The price of Moose has fallen to .06, but even this is thought to be more than it is worth." This time, the Moose estate went to public auction in New York City on January 6, 1883. Francis H. Weeks paid $10,000 for it.[34]

Chapter Thirteen

The Hall and Brunk
Silver Mining Company

I nevitably, the glory days had to end at the Dolly Varden Mine, too, for a mine is a wasting asset. No one can put ore into the earth as a mine takes it out ton by ton. The story of the Dolly Varden, however, contrasts with that of the Moose and illustrates how differently the two mining companies operated.

Assyria Hall and George W. Brunk worked the Dolly Varden without machinery or steam power. The total cost of operation before the mine became productive was $500. Using only hand-drills, Hall and Brunk had focused their work in the Dolly Varden, since the first days in 1872, on slowly prospecting and methodically doing exploration and development work to produce ore. This might not appear noteworthy, except that methodical development was not the norm. Picking out the "eyes of the mine" typified mining strategy. The ore was mined as fast as possible, marketed, and the profits split among the owners. When the ore thinned out, or ended, owners often just walked away, abandoning the mine.[1]

Continuing their methodical approach, Hall and Brunk added twenty-four acres to the Dolly Varden estate in 1877 when they bought the four claims owned by the Park Pool Association that adjoined the Dolly Varden. The Hiawatha, the most important of these claims, had produced $232,799 before Hall and Brunk bought it. In 1875 the Hiawatha had opened and worked out a forty-foot-square chamber of mineral, producing more than 400 tons of ore assaying at 100 to 300 ounces in silver per ton. Even the rock thrown on the Hiawatha dump proved valuable when an assay showed that some of it contained ninety-three ounces in silver per ton. The Hiawatha had kept the Park Pool Association paying dividends: in 1872, a 50 percent dividend; in 1873, 70 percent in cash; in 1874, 75 percent in cash and stock;

and in 1875, after increasing its capital stock from $40,000 to $100,000, the association paid 30 percent in cash.[2]

The press derided Hall and Brunk for not mining ore faster and thereby making the Dolly Varden a bigger, wealthier mine. In May 1877, perhaps as a result of such criticism, they incorporated as the Hall and Brunk Silver Mining Company and seemed ready to follow the same path as the Moose Mining Company. Hall and Brunk's company had a capital stock of $1,000,000, with offices at 85 Liberty Street in New York City.[3]

Hall and Brunk printed an expensive color prospectus that today gives us a snapshot of the Dolly Varden and which, at the time, underlined how difficult the local ore market made Park County mining. They paid an average of ninety dollars per ton to have their ores smelted. The Boston and Colorado Alma Smelter in May 1877 paid only $586 per ton for first-class ore running 630 ounces in silver (worth $756). This $170 difference between the value of the silver in the ore and what the smelter paid the mine represented 22 percent of the ore's value and resulted in an overall cost to Hall and Brunk of $260. Low-grade ores brought even higher charges. The smelter only paid $166 per ton for third-class ore of 205 ounces (worth $246). This equaled 33 percent in charges. For one season the Dolly Varden sold ore to the St. Louis Smelting Company, but the higher transportation costs ate up the higher price that the mine received. This situation led the Hall and Brunk Silver Mining Company to consider opting out of the Park County ore market by constructing its own fifty-ton mill. A mill of their own, Hall and Brunk projected, could concentrate ores for twelve dollars per ton. Then the concentrates would be shipped to Denver.[4]

As they incorporated the Hall and Brunk Silver Mining Company, the two men also incorporated the Mount Bross Tunnel and Mining Company. This company had the grandiose objective of driving a tunnel 8,500 feet, from the area of Dudley, under the Dolly Varden and Hiawatha mines. Then a shaft would be sunk 2,500 feet from the Dolly Varden, connecting the Dolly Varden and Hiawatha to the head of the tunnel. To round out this big plan, they would position the fifty-ton mill at the entrance to the tunnel.[5]

The Mount Bross tunnel would reduce expenses in the Dolly Varden, but it also had a huge front-end cost estimated at $500,000. The richest ore in the Dolly Varden followed a large porphyry dike vertically into the mountain. Hall and Brunk must have considered this dike to be the ore channel for all the silver deposits in that area of Mount Bross. And they must have thought that the Mount Bross tunnel could intersect the porphyry dike 2,500 feet below the mine. This tunnel was a chancy proposition that smelled like another stock scheme, but Charles S. Richardson signed on with the project. He wrote that after about 5,000 feet, the tunnel would cut a large porphyry dike where "surface indications point out a large number of veins; many of these will coalesce, and by their concentration form large, strong, masterly lodes in depth."[6] It seems strange that Richardson used the term "veins" to describe the silver on Mount Bross. But, then, the tunnel would be thousands of feet below the blue limestone and Lincoln porphyry contact and aimed in search of the ore shoot that had created the contact deposits. Tantamount to searching for a needle in a haystack, and, as bait, implying that it would crosscut every vein in the region, the Mount Bross Tunnel and Mining Company did appear on the New York Mining Stock Exchange in May 1878. The stock, however, vanished within a year, and three years later the tunnel was sold at a sheriff's sale.[7]

Unlike the Moose, the Hall and Brunk Silver Mining Company and the Dolly Varden never appeared on the Exchange. Leadville was mining's rising star in Colorado. As that town pulled venture capital into the state, Hall and Brunk concluded (given their experience with the Mount Bross Tunnel and Mining Company) that the possible benefits of the Exchange did not outweigh the profits they were already making in the Park County ore market—at least as long as the Dolly Varden's rich ore held out. They consequently kept the Hall and Brunk Silver Mining Company a privately held corporation.

Their logic made sense as the ore in the Dolly Varden became richer. During the year ending June 27, 1878, the mine returned $80,624, of which $63,164 (78 percent) was clear profit. Between June 5 and 26, 1879, the mine, employing twenty men, shipped thirty-two tons of ore. The first-class ore ran 640 ounces in silver per ton; the second-class, 382 ounces in silver; and the third-class, 239 ounces. This was the richest ore in bulk to have come out of the Dolly Varden. During 1879 the tonnage shipped from the Dolly Varden increased. The workforce grew to thirty men. In September 1879 the mine sold $16,147.50 in ore. After deducting $3,526.33 in expenses, this left a net profit of $12,621.17. About 78 percent of all the ore coming out of the Dolly Varden continued to be net profit.[8]

Hall and Brunk had a bonanza on their hands, and they went straight about their business. The *Rocky Mountain News* published a description of the Dolly Varden in June 1880. Initially, the mine had been worked through Tunnel No. 1. Then the principal access was moved down the mountain a short distance to a second level, Tunnel No. 2, which opened right into the ore house. That tunnel had opened upwards of 400 feet of ore when the

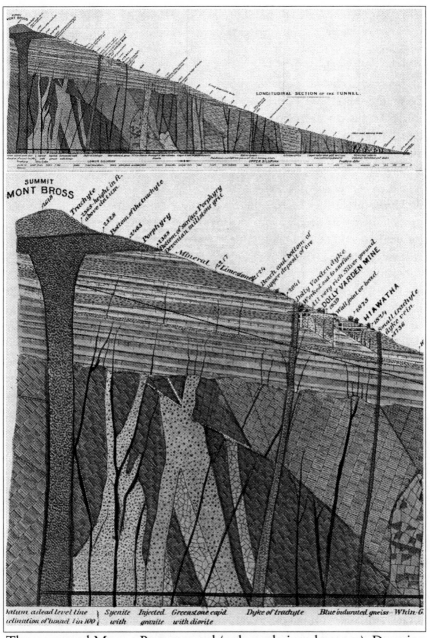

The proposed Mount Bross tunnel (enlarged view, bottom). Drawing by Charles Richardson, Hall & Brunk Silver Mining Company prospectus 1877. Denver Public Library, Western History Department.

work stopped in order to proceed with development in other parts of the mine. A third level, about fifty feet lower, was connected by a shaft. A fourth level had a drift exposing the ore that showed continuously all the way down from the second level.[9]

The ore body in the mine appeared to be in layers, or strata, dipping down the mountain leading to a principal ore body twenty to thirty feet thick. At this point, the miners started a 130-foot shaft down from Tunnel No. 2. In places, ore bodies forty and fifty feet thick had been left untouched. Well-defined layers, or floors of ore, had never been worked. By 1880 the distance into the mine from the main tunnel had become so great that Hall and Brunk needed to drive another tunnel (the Hiawatha tunnel) farther down the mountain to strike the main shaft at the third level. The Hiawatha tunnel would allow the miners to output the bulk of the ore without the expense and delay of handling it several times. On the surface, the main entrance into the Dolly Varden overlooked a dump, where thousands of tons of ore had been thrown. This ore had been too low-grade to be profitable in the Park County ore market, but in one part of the dump, the ore reportedly ran 218 ounces in silver to the ton. The miners, without sufficient room for ore sorting, finally had to leave most of the lower-grade mineral (between fifty and sixty ounces in silver per ton) standing in the mine.[10]

From 1872 to 1880 the mine produced $443,474 in ore, making the total production from what became the Dolly Varden estate $684,273. Hall and Brunk had made a profit of $136,008 (31 percent) in the Dolly Varden, and the profit would have been even larger had they not reinvested their money by purchasing adjoining property, procuring patents for their claims, running tunnels, sinking shafts, and erecting buildings. The average cost of mining was forty-five dollars per ton. Hauling the ore down Mount Bross cost four dollars per ton, and then it was sold in the noncompetitive Park County ore market. Hall and Brunk felt that there was no possibility of losing the ore chute in the Dolly Varden for years to come.[11]

Early in the winter of 1880–81, George Brunk kept the miners doing development work. He stated that he did not intend to break any ore until spring. But suddenly, in February 1881, the miners began removing ore so fast that jack trains made two trips a day to keep the ore house clear. This spurt of activity signaled a change in plans. Hall and Brunk had entered into a sales agreement to sell the entire Dolly Varden estate, and they were going to output ore until the property left their hands.[12]

The purchaser, the Boston Gold and Silver Mining Company, organized in 1880, operated near Breckenridge in Summit County where it owned lodes, placers, and a small mountain smelter. On February 21, 1881, Hall and Brunk led Edwards H. Goff, of Boston and Montreal, who headed the company, and Prof. J. Alden Smith, Colorado State Geologist,[13] on a tour of the Dolly Varden. Upon returning to Hall and Brunk's office in Alma, they concluded the purchase of the Dolly Varden for $400,000. Jubilantly, Park County anticipated that more investment capital would follow this sale.[14]

The Boston Gold and Silver Mining Company soon announced a two-pronged strategy with the Dolly Varden estate. First, the company intended to upgrade the road over Hoosier Pass for year-round use so that the high-grade ore from the Dolly Varden could be hauled straight to its Boston and Breckenridge Smelting Works. Second, the company wanted to process the extensive mine dumps at the Dolly Varden. Estimates of the ore in these dumps ranged from 12,000 to 20,000 tons. If this ore averaged about thirty ounces of silver per ton, the smelting works could treat it for ten dollars per ton, leaving a profit of $240,000, minus the cost of handling and hauling.[15]

Brunk waited for the new owners to pay the purchase price and send a manager to take charge. By the end of February, with no change of ownership confirmed, Brunk started a drift north from old mine workings into ground that had never been prospected. After twenty-five feet, the miners struck a body of mineral assaying 110 ounces in silver per ton. The ore body extended about 500 feet west from the opening, and 300 feet east. Brunk also resumed work on the Hiawatha Mine, where a new shaft encountered a deposit of ore three feet thick. The Dolly Varden just kept getting richer.[16]

The owners of the Boston Gold and Silver Mining Company launched a vigorous advertising campaign to promote a new stock offering based on the Dolly Varden purchase. But they shot themselves in the foot when they stated that the Dolly Varden and Moose were at that time the richest mines in Colorado. Many people knew better. The Moose Mining Company, only a shadow of what it had been, verged on collapse. The first public indication of problems with the Dolly Varden sale came early in July 1881, when the *Denver Republican* rumored that the deal had fallen through. The newspaper carped that "Banker Goff, of Boston, must have expended quite a sum of money learning that there are instances where advertising does not aid success in mining."[17]

The Boston Gold and Silver Mining Company did, however, incorporate the Dolly Varden estate into its holdings, on paper at least, offering a capital stock of $2,000,000 divided into 200,000 shares of ten dollars each. But by late August, Hall and Brunk were still in possession of the Dolly Varden, waiting for the new owners to send a manager. All the geological reports on the Dolly Varden estate had proved favorable, but, interestingly, sections of the reports published in the *Mining Record* and the *Engineering and Mining Journal* implied that little high-grade ore was in sight. The Dolly Varden estate's value lay in its immense quantities of lower-grade silver ore, both in the mine and on the dumps.[18]

Would handling, transporting, and smelting low-grade ore from the Dolly Varden pay? A mill run of 15,000 pounds of ore from the dump did yield twenty-five ounces in silver per ton, but this sale came to naught—not because of the Dolly Varden, but because the Boston Gold and Silver Mining Company's pyramid of Summit County mining schemes collapsed. After paying ten dollars in fees in April 1882, Hall and Brunk got the entire Dolly Varden estate back.[19]

Following the completion of the Hiawatha tunnel to the main shaft came reports of "wonderfully rich ore" from the Dolly Varden during 1882. The

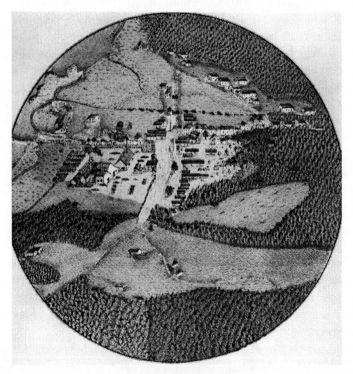

View of Alma from Dolly Varden Mine. This drawing by Charles S. Richardson is in the 1877 Hall and Brunk Silver Mining Company prospectus. Denver Public Library, Western History Department.

The crew of the Dolly Varden Mine. Mazzula Collection, Amon Carter Museum.

production reached eighty tons per month. Even the ore market improved a bit when the Boston and Colorado Sampling Works increased its buying prices. Ore carrying only thirty ounces in silver (at $1.14 per ounce) now left a profit of $2.80 per ton after deducting for loss and shipment to Denver. Still, Park County did not offer an ore market for the mine dumps at the Dolly Varden.[20]

Things at the mine took a turn for the worse on September 17, 1882, when the balance of the miners were paid off and discharged. The workforce was reduced to five men, the smallest it had been in years. Hall and Brunk also incurred the first debt against their mining property when they signed a promissory note for $25,000 on September 29. This gave them working capital, but Henry R. Wolcott and Nathaniel P. Hill of the Boston and Colorado Smelting Company now held a deed of trust for the Dolly Varden estate. The glory days were obviously over.[21]

Early in 1884, Hall and Brunk made repeated tests of the ore in the mine dumps. Sampling ore in amounts this large was a challenge, but they concluded that the estimated 25,000 tons in the dumps averaged thirty-five dollars per ton ($875,000) in silver. Hall and Brunk let a contract to haul 1,000 tons from the mine dumps to the Boston and Colorado sampler at Alma in July 1884. They planned to follow this with shipments of forty to fifty tons per day, but nothing was said about the outcome of the first shipment of 1,000 tons.[22]

Both the Moose Mining Company and the Hall and Brunk Silver Mining Company had incorporated for twenty years. Rarely did any mining company live out the term of its incorporation. On July 12, 1897, the stockholders of the Hall and Brunk Silver Mining Company convened at the company's office in Alma. All the stock in the company was represented. George W. Brunk, president, owned 500,000 shares. Assyria Hall, vice president, owned the other 500,000. In a corporate meeting combining humor and reflection, Brunk acted as presiding officer and Hall as secretary. Hall presented a resolution to cancel the promissory note by transferring the entire Dolly Varden estate to Crawford Hill, heir of Nathaniel P. Hill, for the sum of one dollar. Almost twenty-five years to the day after their discovery, these two men, who had prospected the ore shoot on the Dolly Varden, measured and marked the boundary lines, filed the claim, incorporated the mine in 1877, and then worked the mine as a bootstrap operation, disbanded the Hall and Brunk Silver Mining Company.[23]

Chapter Fourteen

Living at
These High Mines

No miners anywhere in North America worked at higher altitude than those on Mount Bross and Mount Lincoln. Perched on slopes as steep as sixty degrees at over 13,000 feet, these mines sat on little level ground. When the men had time and when the weather was pleasant enough to be outside, the only recreational activity available to them was either hiking up, or hiking down. One word summarizes what it was like to live at these high mines: confining.

The weather could be extreme and unpredictable.[1] No comprehensive account of the weather conditions on Bross and Lincoln exists, but accounts from various sources give some idea of the challenges the miners faced living in such precarious quarters.

The first glimpses come from tourists. Following the isolation of winter, summer brought visitors to the mines. Several parties had climbed Mount Lincoln in the 1860s before silver had been discovered, but the completion of roads in 1873 from the Moose Mine to the Present Help Mine, and from Quartzville up the southeast spur of Lincoln, made these mountains easily accessible.[2]

In the early 1870s Mount Lincoln was thought to be the highest mountain in Colorado. Named in honor of President Abraham Lincoln, and even more importantly, because it had two roads up it, Lincoln started receiving national attention due to tourism. And, indirectly, the mining received attention, too. On August 31, 1873, a touring party composed mainly of women rode saddle horses to the Present Help. The group was on a tour of the West, and Grace Greenwood, one of the group's members, sent accounts of what they saw to the *New York Times*. As they rode up Mount Bross, she described:

...curious cabins, perched on the steep and crumbling mountain side, like swallow's nests against the rocks. Their roofs, like those of Swiss chalets, were kept in place by heavy stones. If I were a miner I should prefer to lodge in my tunnel—I would rather creep into the bosom of mother earth—than hang on to her ragged skirts when the winds were rising and avalanches falling.[3]

When the group reached the small boarding house covering the entrance to the Present Help, Eliza Greatorex, another member of the group, wrote:

The wood which they burn costs $60 per cord; but they have good food, excellent beef, and the bread which we saw the handsome young baker putting in the oven was of the most encouragingly light appearance.[4]

While the location of the Present Help at 14,157 feet on Mount Lincoln would seem decidedly out of the way, another party of tourists also visited on August 31, 1873. This group included Anna E. Dickinson (1842–1932), who had attained national fame by speaking out for abolition and women's suffrage. Dickinson had come to Colorado to restore her health in the mountain air and, in the process, developed a passion for mountain climbing. Greatorex mentioned Dickinson and her party of tourists:

The miners gave us a most friendly invitation to stay and share their meal. They told us that Anna Dickinson and her party had dined with them. There were seventeen men together, and their trim house, with everything, beds, cooking arrangements, and fair-spread dinner-table, gave one the idea of a well-manned ship at sea. Stopping to look back, after we had left their house, to get a sketch of it, we saw that the miners had gathered round Miss Dickinson, who had come back to say good-bye to them. They stood in honestly-expressed admiration of her pretty, gay costume, and of her bright face. They had, while our party were there, been telling one another how famously she and Grace Greenwood were to "write up" their mines.[5]

By July 1874, when author Helen Hunt Jackson came to Fairplay from Colorado Springs, carriages could be hired in Fairplay for the trip up Lincoln. After an early morning start, the horses' sides heaved like bellows and their breathing was loud as they went up Mount Bross. On the steep climb to the Dolly Varden, the horses could go only about two or three times the length of the carriage before they had to stop to rest for five or six minutes. Helen Hunt Jackson thought the buildings at the Dolly Varden Mine projected like odd-shaped rocks from the side of the mountain. Some of the miners stood in the cabin doors as her carriage crept by. "I gazed at them earnestly," she wrote, "expecting to see them look like sons of gnomes of the upper and lower air; but their faces were fresh, healthful, and kindly."[6] Shortly afterward, her carriage passed over the Moose Mine, crossed the northeast face of Bross to the saddle with Mount Cameron, and then on to Lincoln.

A fierce wind blew when Jackson arrived at the welcome roaring fire ablaze in the boarding house at the Present Help. She also remarked that the miners' quarters resembled a ship with the bunks ranged in tiers, and she wrote this about the mine:

> The cook was a cheery fellow, with a fine head and laughing brown eyes. He was kneading bread. His tin pans shone like a dairymaid's. The cabin was by no means a comfortless place. One wide, long bench for table; a narrow one for chairs; tin cups, tin pans, black knives and forks,—we borrowed them all. The cook made delicious coffee for us and we took our lunch with as good relish as if we had been born miners. The men's beds were in tiers of bunks on two sides of the cabin, much wider and more comfortable than stateroom berths in steamers. In each berth was a small wooden box, nailed on the wall, for a sort of cupboard or bureau drawer. In these lay the Sunday clothes, white shirts, and so forth, neatly folded. There were newspapers laying about, and when I asked the cook if he liked living there, he answered: "Oh, yes! very well. We have a mail once a week." A reply which at once revealed the man was significant of the age in which he lived.[7]

It took Jackson five hours to travel the twelve miles from Fairplay to the Present Help, and after she climbed the short distance to Lincoln's summit, her carriage had her back in Fairplay by sunset.

One other description of an ascent of Mount Lincoln deserves mention. In August 1876 a Mrs. Cameron and Mrs. Wirtz, starting from Alma, hiked up the Mount Bross road. They reached the Present Help Mine (six miles from Alma) early in the afternoon, where they spent the night. The next morning they ascended Lincoln, descended the east side of the summit to the Russia Mine, and then hiked down through Quartzville and back to Alma. "Both are ladies of mature age and considerable avoirdupois," the *Denver Mirror* said. That was quite a hike.[8]

Some glimpses of the weather come from Charles S. Richardson, who jotted down descriptions from the winter of 1875–76 in his survey notebook during his stay at the Moose Mine. He noted "heavy gale winds," with the temperature at minus two degrees at 7 A.M. on the morning of October 25, 1875. In contrast, on the morning of January 12, 1876, he wrote, "This is the most beautiful morning I have ever seen on this mountain. Temperature 24 degrees, no wind. Bright sunshine—lovely."[9] Actually, that winter turned out to be unusual. Very little snow fell in Alma, with mild temperatures prevailing for weeks.

How many winters anyone tried to stick it out at the Present Help Mine on Mount Lincoln is not clear, but a miner there reported in early February 1876 that the weather had been so pleasant for the past three weeks that he could work outdoors (at 14,157 feet) in shirtsleeves and without gloves. These warm temperatures also produced widespread attacks of pneumonia, flu, and other winter crud, which flourished in the crowded and confined sleeping rooms at the Moose, Dolly Varden, and Russia. The greater part of the large force of men working at the Moose was under the weather for a week.[10]

Possibly a more typical early morning temperature was one recorded at minus thirty-five degrees in Alma at 6 A.M. on November 17, 1880. It would have been even colder 3,500 feet higher at the Moose Mine. Whether it was the temperature or close quarters and working conditions that strained relations and led to a strike at the Russia Mine is not clear. But in 1879 the Russia hired two men whom, by reason of their nationality, the other miners found objectionable. The miners stopped work, demanding that the new

Morning clouds cover Hoosier Pass from Moose Mine; Grays Peak looms at right. Charles S. Richardson Collection, Notebook 26, Denver Public Library, Western History Department.

arrivals be discharged. The foreman, who had hired the men, refused to dismiss them. But the superintendent, with his eye on production, sided with the striking miners. The foreman resigned, and the two objectionable workers were let go.[11]

The miners did not "lodge" in their tunnels as Grace Greenwood suggested, but all the mine buildings on Mounts Bross and Lincoln did huddle up against the mountainsides to get protection from the elements and preserve heat in the living quarters. At the Dolly Varden, the main door led directly into the ore-sorting room, which connected into the mine. A number of men sorted the different grades of ore here in a thirty-by-sixty-foot room with large ore bins at the side walls. In one corner stood a huge chest containing supplies needed by the miners underground. In another corner was a blacksmith forge and furnace. A hallway led from the ore-sorting room to a comfortably furnished sitting room for the men. Adjoining this was the long sleeping room fitted with wide, roomy bunks. Next came a storeroom for canned fruits, coffee, syrups, flour, and so forth. The mine foreman occupied the room beyond the storeroom, and this was followed by the dining room, which was well lighted and had tables long enough to seat twenty-four. After the dining room came the kitchen, and then another storeroom for the beef that was brought up to the mine. This building was 125 feet long. The men could go through the entire building, through the ore-sorting room, and into the mine without ever going outside.[12]

How the Dolly Varden building was heated in the winter is not known. Hauling large quantities of wood or coal would have been costly, and since insulation was not readily available, everyone probably dressed warm in the winter when they were not underground. Staying inside obviously prevented exposure to the intense cold in the winter, but more important, the air temperature averaged nearly forty degrees underground in the mines. Thus, the mines offered escape from the intense winter cold. But some activities, such as hauling ore down and supplies up, had to be done outside. This was a never-ending chore. When the ore wagons, or in winter the jack trains, came up to the mines, they hauled supplies. They then took ore back down. The frequency of this traffic kept the mines supplied, but occasionally winter blizzards halted traffic on Mount Bross for up to a week. One snowfall, approaching thirty-six inches, fell on April 26, 1877. A traveler reported that the snow was so deep that the tops of the twelve-foot-high telegraph poles from Alma to the Moose Mine were good places to sit and rest during the ascent of Bross. It was not until July 22, 1877, that the road to the Moose was cleared to allow teams and wagons to get to the mine.[13]

The *Engineering and Mining Journal* in December 1877 described a trip from Dudley to the Moose Mine on a cold and windy day:

The ascent of Mount Lincoln [Mount Bross] in the winter is an undertaking of some labor. One has only to look upwards at the great white mass, over which the wind sweeps in continuous gusts, raising great whirls of snow, which carom along the steep slope for hundreds of yards at a time before they are swallowed up by a deep ravine, or dissipated on a tall crag. Can one mine up there in the winter? Alma is 10,000 feet above the sea. There is only a narrow fringe of timber, 1,000 feet wide above it, and then comes 3,000 perpendicular feet of bare rock, so steep that the loose rock will scarcely cling to its side. But a glance through a glass, at almost any time of the winter, will reveal a train of hardy "burros" climbing up to the silver mines, or returning heavily laden with ore, or a horse-man, or pedestrian, beating his way against the fierce storms. There is no better proof of the value of these mines than the energy with which they are worked.

The road leads up through Alma, along the base of the hill to Dudley, where are the offices and reduction works of the Moose Mine. A little above here the path turns abruptly to the left, and you begin the ascent. So long as timber holds out the way is easy. You follow a well-graded wagon road, and experience considerable comfort in the thought. The pine and spruce rapidly thin out, however, as the brow of the first bench is reached...progress is slow. It is a weary pull. The way leads up, up across dazzling fields of snow set on edge, across wind-blown patches of limestone chips, where the foot slides at every step, still upwards in the face of the blast loaded with a million little ice crystals to the square inch which would almost shave a man if permitted, till at last a camp of half buried shanties is reached which mark the openings of the Dolly Varden Mine. Everything is closed up tight. Not a worker is to be seen. They are all hidden away in the heart of the mountain, and though the appearance of human habitations apparently so untenanted is somewhat dispiriting, yet it is otherwise encouraging to the traveler, for he knows that at last the edge of the metaliferous limestone is reached. For the Dolly Varden Mine is on the same belt as the Moose.

Winding along upon the slope, the road at last takes a turn to the left around a sharp point. Here the wind is at its best, and bowls its way against you at any rate which a lively imagination or an anemometer can suggest. It is a short pull, however. Another collection of snow-bound buildings appears snugly set into the mountain side, on the edge of a deep ravine, and in a minute after a final struggle with the blinding storm you find shelter in Superintendent Hill's office, when the consoling fact is learned that the thermometer outside stands at 18 degrees below zero, and the wind gauge records 40 miles an hour.[14]

One benefit of the cold temperatures and permanently frozen ground was that the Moose stayed dry underground. The only timber needed to support the mine's walls or roof was where the tunnels first penetrated the surface slide rock. Like those at the Dolly Varden, the buildings for lodging and feeding the miners were built at the mouth of the mine. The buildings at the Moose stretched along the mountainside, allowing the miners to step from their sleeping rooms and mess hall directly into the mine. All the sorting and sacking of the ore was done underground. The miners loaded the jack trains in a long shed connected to the mine's entrance before the burros ventured out and down Mount Bross.[15]

Situated on the northeast slope of Bross, the Moose Mine buildings were reached by little sunlight during the winter. The men lived lives devoid of all outdoor enjoyment. After their shifts, they generally congregated to play cards or checkers in either the long, low dining room, in the sitting room

Collapsed structure at location of Present Help Mine, with Mount Lincoln summit at right. Photo Harvey Gardiner.

adjoining it, or in the bunkroom. Superintendent Hill gave all the men free access to his respectable library of well-selected books. Only a telegraph line to Dudley kept the mine from total isolation.[16]

Shortly after the *Fairplay Flume* started publishing, the editors visited the Moose on a windy day early in March 1879. After climbing "up the sides of a mountain so steep that every step is like climbing a ladder," the *Flume* wrote, "one is well repaid for their tramp by the glimpse which is caught of the manner of working and living on an enormous mine at the great altitude of the Moose, and one is forced into an expression of admiration for the pluck of these men."[17] On this particular evening a quartet or more of men gave some "fine vocal music," and the "semi-occasional passage of a bottle" also put the men in high spirits as the concert progressed until a very late hour.

In addition to struggling with these extreme weather conditions and confined living quarters, the miners found little joy in their daily meals. The menus at these mines were of necessity repetitious, and the miners at the Dolly Varden periodically complained about the food. On Saturday, August 1, 1880, in the warmth of summer, seventeen miners from the Dolly Varden marched into Alma on strike. It was the old trouble with the cook, they said. George Brunk, however, learned that one or two of the instigators had compelled the other men to join in the "strike," and he had the ringleaders arrested. The other men returned to the mine the next day. It seems that about every two months the men at the Dolly Varden went on "strike" and walked down to Alma for the usual "gamboree." Either the menus or the cook at the Present Help was exceptional, or the women visitors who had enjoyed the food there received special fare.[18]

The number of men working on Bross and Lincoln changed constantly. The census of 1880 counted ninety-nine men when the census takers came through in the summer. The Moose served as Park County Precinct No. 17 in the elections of 1880 and 1882.[19]

How often the men came down from these high mines is not clear, but their "gamborees" often made it hard for them to get back up. Late on the Saturday afternoon of January 24, 1885, two Dolly Varden miners left Alma. Both were pretty drunk, and the night turned terribly cold before they reached the mine. When the first arrived, he could give no account of what had become of his companion. A rescue party went out but could not find the other man that night. It was feared that he had frozen to death, but on Sunday he was found in a cabin not far below the mine, having been sober enough to save himself. His horse had the good sense to enter a shaft house nearby and save its life, too.[20]

While nearly all of the miners longed for any opportunity to get off the mountains, a few isolated themselves voluntarily. A miner named Fred Pitman at the Dolly Varden went for a stretch of 140 days without leaving the mine. He said that at another time he had not left a mine on Mount Lincoln for 290 days. The longest record in this category, however, went to John S. Borders. Borders gradually consolidated a number of claims, known as the Danville Group, on the northeast spur of Mount Lincoln. He went up there heavily in debt, vowing not to come down until he was free from his debts. After 395 days, he came down from the mountain several thousand dollars ahead of the game.[21]

Chapter Fifteen

The Power of the Metals Market

For two decades following 1871, the price of silver gradually trended downward. Blame lay with the prodigious success of mining, which between 1850 and 1875 increased the world's production of gold and silver extraordinarily. Greater stocks of gold encouraged nations to substitute gold for silver in their currencies. Several European nations had done this by 1873. The United States stopped coinage of silver dollars in what is called the "Crime of 1873," thereafter using silver only for fractional coinage.

Nations displacing silver currency with gold caused large amounts of silver to be thrown upon the international market. The ever-increasing output of silver mines—Park County and Leadville being two of the contributors—only added to the surplus. India and other Asian countries relieved some of the downward pressure on the price of silver by continuing to base their currencies on silver. They absorbed the excess silver for a number of years.

Domestically, this complex struggle between gold and silver pitted West against East in the United States. Western silver-mining states demanded "free coinage" of silver at the ratio of sixteen ounces of silver to one ounce of gold. Eastern interests felt that heavy government purchases of silver would replace the gold dollar with a depreciated silver dollar or, even worse, with another issue of greenbacks (paper money) as had been done during the Civil War.

On two occasions—in the Bland-Allison Act of 1878, which restored the silver dollar as legal tender, and in the Sherman Act of 1890—Congress authorized the purchase of silver for coinage into silver dollars at the U.S. mints. But the production of silver still exceeded the needs of the mints, the needs of the international market, and the needs of private business.

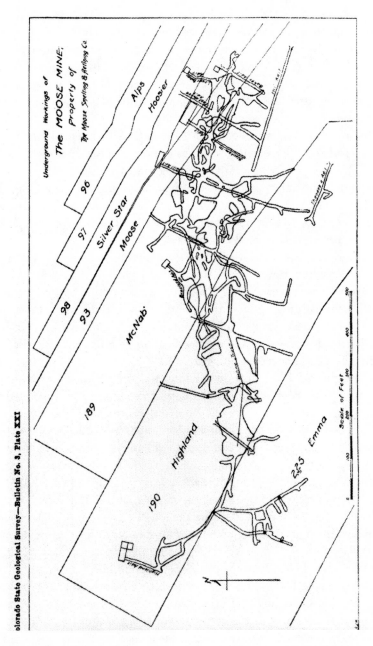

Moose underground workings, 1911. *Colorado Geological Survey Bulletin #3* (1912).

The United States did mint millions of silver dollars, but in the early 1890s banks began hoarding their gold by paying receipts owed to the government in silver. Government, in the course of its financial activities, paid for purchases in gold. As millions of silver dollars sat in storage at the mints, silver was slowly replacing the gold reserves of the United States Treasury. The possibility that the United States might suspend gold payments caused great concern in certain quarters.

In the twenty-two years between 1871 and 1893, the price of silver had slowly eroded 37 percent from $1.32 per ounce to $.83 per ounce in June 1893. Business in western silver-mining states began retrenching amid feelings of financial apprehension. When India announced in June 1893 that it would cease the coinage of silver, the international silver market started tumbling. In four days' time the New York price quote for silver fell 25 percent from $.83 to $.62 per ounce.

The crash of silver floored the silver-mining industry. Silver mining halted everywhere in Colorado as mines and smelters ceased activity until a recovery in the price of silver would permit profitable operations. At the mines, men just quit, walked away, and left everything as it stood. When the financial panic broke full fury in July 1893, twelve Denver banks closed their doors within three days. Businesses failed, and thousands of workers and miners were out of work.

The Panic of 1893 was an obvious watershed for silver mining. In a major mining center like Leadville, whether a mine was isolated with bonanza-grade ore, or productive but not particularly efficient, or well developed, productive, and efficient, silver mines stopped working when the price of silver collapsed. The metals market had convincingly demonstrated its power.

The Moose Mine's glory days were long gone by 1893, but in those troubled financial times the Moose briefly reappeared in the high financial circles of New York City. Following the failure of the Moose Silver Mining Company, the bondholders, who faced losing everything, agreed to put up $10,000 to buy the entire Moose estate at a public auction in New York City in January 1883. By doing this, they hoped to find a buyer and to make a future sale possible. Francis H. Weeks, one of the organizers of the Moose Silver Mining Company and now sole trustee for the Moose estate, was chosen as the man with the most potential for selling the Moose, and, accordingly, the deed was drawn up in his name.[1]

Weeks had, since the early 1880s, been a full-fledged hustler investing in land improvement schemes. He acted as trustee for a number of sizeable

Ludlow Patton. Photo by Henry Whittemore, *The Founders and Builders of Oranges* (New Jersey: L.J. Hardharm, 1896).

estates and, by managing his trust interests well, had earned a reputation as a safe person to entrust with funds to earn high interest. Potential investors besieged his office, anxious to make him custodian of their money solely on a guarantee of good faith.[2]

Ludlow Patton (1825–1906),[3] a wealthy former member of the New York Stock Exchange and Mining Stock Exchange and owner of the majority of the defunct Moose Silver Mining Company bonds, agreed to advance money to Weeks to pay the taxes and expenses for the care and custody of the property. This constituted a first lien, with Patton to be repaid from any eventual sale of the Moose estate. For ten years Patton kept advancing Weeks money, but Weeks never found a buyer.

In reality, Weeks had a pyramid of financial schemes under way, some of which proved profitless, and he siphoned more than $600,000 from his trust interests, mostly widows and children. Then the uncertain economic times unraveled his schemes in May 1893 and Weeks had to make an "assignment" to his creditors, meaning that he had to deed over his assets; the Moose estate was an insignificant piece of this very large whole. Weeks fled to Costa Rica with his wife. He was extradited back to the United States and got ten years' hard labor at Sing Sing Prison.[4]

Compared with the widows and children who discovered that Weeks had embezzled their estates, the Moose estate fared well because by that time it was nothing more than abandoned mining claims in Colorado. This did put Ludlow Patton to considerable trouble, however, because he had to take the matter to the New York State Supreme Court to regain control of the estate.[5]

Another attempt at reviving mining at the Moose came after Patton's death. His second wife, Marion, had remarried to Henry W. Scott. In 1910, when the price of silver was as low as $.54 per ounce, the Scotts journeyed to Colorado to see the Moose Mine for themselves. Evidently mesmerized by high-altitude dreams of glittering wealth, they decided to reopen the Moose estate. This proved to be a fatal miscalculation that cost them many thousands of dollars and ultimately destroyed their marriage.

As recently as 1980–81, as the price of silver gyrated wildly between eight and fifty dollars an ounce before declining to ten dollars an ounce, the Moose estate showed that it could still pay if the price of silver was right. Miners working on a large body of low-grade ore in No. 9 Tunnel on the Highland claim produced 40,444 ounces of silver worth $430,508.[6]

Epilogue

Many Colorado silver mines produced more wealth than the Moose, but the Moose is easily the most important silver mine in the state's early history. Plummer and Myers's discovery of the Moose in 1871 represents the signal mineral discovery of silver in a horizontal blue limestone and porphyry contact. Their find, and the excitement that followed, is a clear example of how one mineral discovery led other prospectors and mining men step by step to further discoveries, and ultimately to the blue limestone and porphyry contact at present-day Leadville.

The Moose contained the largest and richest outcrop of the contact on either side of the Mosquito Range. In contrast, the blue limestone contact in the Leadville area appeared in only a few outcrops due to widespread covering by glacial deposits and "wash."

On the Park County side of the Mosquito Range, William H. Stevens developed a critical eye for the blue limestone and porphyry contact outcrops. From that experience on Mounts Bross and Lincoln, he came to understand the geological significance of silver in horizontal deposits of blue limestone. On the Lake County side of the range, Stevens surmised that the heavy, dark sand plaguing the California Gulch placer miners was eroded from lead silver carbonate ore. When he traced the ore to its source in horizontal outcroppings of the blue limestone and porphyry contact at Leadville, he could speculate on the richness of the silver deposits to be found there.

Harvey Gardiner stands above the location of the Moose Mine, with Mount Silverheels (13,817 feet) in the background. Photo Phyllis Hunt.

Notes

Abbreviations

CHS: Colorado Historical Society
CT: *Colorado Tribune*
DM: *Denver Mirror*
DPL: Denver Public Library,
    Western History Department
DR: *Denver Republican*
DT: *Denver Times*
DTB: *Denver Tribune*
EMJ: *Engineering and Mining Journal*
FF: *Fairplay Flume*
MR: *Mining Record*
MRV: *Mining Review*
NYT: *New York Times*
RMN: *Rocky Mountain News*

Chapter 1

1
Theodore Francis Van Wagenen Papers, "Notes Colorado," DPL.

2
Daily Highest and Lowest Prices of Gold at New York for the Four Years from Jan. 1, 1862 to Jan. 1, 1866 (N.Y.: Commercial and Financial Chronicle, n.d.).

3
Frank Hall, *History of the State of Colorado* (Chicago: Blakely Printing Co., 1889), 1: 306–9.

4
"Some Records of the Early Days," *MRV* 4 (June 1874): 38–39. "Some Records of the Early Days N.2," *MRV* 4 (August 1874): 52–53. "The Profits of Gold and Silver Mining," *MRV* 6 (August 1875): 119–20. "Investments in Mining," *MRV* 8 (May 1, 1876):

264–65. "Mining," *RMN,* January 11, 1877; "Old Colorado," *EMJ* 23 (May 26, 1877): 353.

Chapter 2

1
"From Our Traveling Correspondent," *CT,* August 7, 1870.

2
William H. Brewer, *Rocky Mountain Letters 1869* (Denver: Colorado Mountain Club, 1930), 28. "From Montgomery District," *RMN,* October 26, 1861; "From Our Traveling Correspondent," *CT,* August 6, 1870; "Park County," *Georgetown Colorado Miner,* August 11, 1870.

3
"From Our Traveling Correspondent," *CT,* August 7, 1870.

4
"From Our Traveling Correspondent," *CT,* July 14, 1870; July 16, 1870; August 3, 1870. "Letter from Nil Desperandum," *CT,* October 20, 1870.

5
"From Montgomery," *RMN,* July 7, 1865. "Park County," *Georgetown Colorado Miner,* September 7, 1871. Joseph H. Myers, born in Pennsylvania on October 31, 1834, came to Colorado on August 8, 1859. Hall, *History of the State of Colorado* 2: 560. Virginia McConnell, *Bayou Salado* (Denver: Sage Books, 1966), 81, 84, 112, states that travelers often stayed with the Myers family and that "Mr. Myers" led them on climbs of Mount Lincoln. A title search shows that the only Myers living at Montgomery in the late 1860s was Joseph H. Myers. For additional information, see chapter 10, note 18.

6

See R.S. Morrison, *Mining Rights in Colorado* (Denver: Chain & Hardy Publishers, 1881), 24–25. Henry N. Copp, *Decisions of the Commissioner of the General Land Office and the Secretary of the Interior* (San Francisco: A.L. Bancroft and Company, 1874), 201. The bylaws of the Consolidated Montgomery District are in the Compromise lode, Patent Case File 1351, BLM Records Group 49, National Archives, Washington, D.C.

7

D.K. Sickels, *The U.S. Mining Laws* (San Francisco: A.L. Bancroft and Company, 1881), 43–44.

8

Dwight lode, Patent Case File 573, BLM Records Group 49, National Archives, Washington, D.C.

9

"Letters to the Editor, Fairplay, August 4, 1871," *RMN*, August 10, 1871. "Letter from Buckskin," *DTB*, September 11, 1871.

10

"Mining Excitement at Fairplay," *DTB*, July 28, 1871.

11

Park County, Book D, 146. "The New Mines of Mt. Bross," *DTB*, August 10, 1871. Three assays of ore from Mount Lincoln ranged from a low of twenty-seven ounces in silver ($35.63 per ton) to a high of ninety-four ounces in silver ($122.83 per ton).

12

History of the City of Denver, Arapahoe County, and Colorado (Chicago: O. L. Baskin and Company, 1880), 409. Richard E. Leach, "Judson H. Dudley,"

The Trail 4 (October 1911): 15–17. "Judson H. Dudley Nears the End of a Long, Active Life," *DR*, February 13, 1900. For additional information concerning Judson H. Dudley, see chapter 10, note 18.

13

Park County, Book F, 364. Andrew W. Gill was a mining broker on Wall Street.

14

Park County, Book D, 168–70.

15

Noted in "Letter from Buckskin," *DTB*, September 11, 1871.

16

This description of the burro is from "Notes Colorado," 15–16, Van Wagenen Papers.

17

"The Mounts Lincoln and Bross Silver Mines," *RMN*, December 13, 1871.

18

See "Park County Correspondence," *Caribou Post*, November 4, 1871, and Frank Fossett, *Colorado: Its Gold and Silver Mines, Farms and Stock Ranges and Health and Pleasure Resorts* (New York: C.G. Crawford, 1880), 504.

Chapter 3

1

The description of the legal requirements embodied in the Mining Law of 1866 is based on Morrison, *Mining in Colorado*, 210–15, and the Moose lode, Patent Case File 546; Dwight lode, Patent Case File 573; Dudley lode, Patent Case File 574; Bross lode, Patent Case File 575, and Gill lode, Patent Case File 576, BLM Records Group 49, National Archives, Washington D.C.

2
"History of Gloversville. From the Gloversville *Daily Leader* of October 28, 1899," 74. James H. Manning, ed., *New York State Men* (Albany: Albany Argus Art Press, 1914), n.p.

3
Park County, Book F, 93.

Chapter 4

1
Park County Bulletin, December 23, 1910. "Colorado '59er Dead," *Rocky Mountain Herald,* December 31, 1910.

2
"Cy Hall, Wealthy Mine Owner, Dies Here at Age of 76," *Monrovia Daily News* (Calif.), June 9, 1917. "Assyria Hall," *The Trail* 10 (August 1917): 30.

3
The Alps was located by Addison M. Janes and thirteen others, Park County, Book D, 149. The Alps was immediately quit-claim deeded to Myers, Hall & Brunk, Park County, Book F, 350. The Hoosier, Book D, 150 and the Hiawatha Book D, 153.

4
Park County, Book D, 151.

5
Park County, Book D, Lincoln, 148–49; Present Help, 171, and Silver Star, 181.

6
"Leadville's Columbus," *DTB,* February 2, 1879. "Man Who Founded Leadville Is Dead," *Denver Post,* March 7, 1901. "Remarkable Life Is Closed," *RMN,* March 8, 1901. "Blunt, but Kind, Capt. Stevens Was One of the Most Peculiar Characters among Detroit's Most Wealthy Citizens," *Detroit Free Press,* March 8, 1901. Iva Evans Morrison, "William H. Stevens," *Colorado Magazine* 21 (July 1944): 121–29. Frank Hall, *History of the State of Colorado* (Chicago: Blakely Printing Co., 1890), 2: 428–32. A lease-permit system for federal lands was used in Michigan.

7
Park County, Book D, 146. The next entry in Book D is the location of the Moose on August 5, 1871.

8
Stevens's fifty-dollar purchases appear in the Park County, Book F, Wilson lode, 380; Hiawatha lode, 381; "William H. Stevens" lode, 382; Haines lode, 383; and Alpine lode, 386. In addition, Stevens bought the following placer claims: forty acres of land in Quartzville Gulch for one dollar, Book F, 375; 160 acres near the Moose and 160 acres of Quartzville Gulch, Book F, 376; and eighty acres "in the center of Summit Gulch" and eighty acres "in the center of Placer Gulch," Book F, 378.

9
The organization of the Park Pool Association is in the James V. Dexter Papers, folder 106, CHS.

10
William N. Byers, *Encyclopedia of Biography of Colorado* (Chicago: Century Publishing & Engraving Company, 1901), 1:231–32. Jerome C. Smiley, *History of Denver* (Denver: J.H. Williamson & Company, 1903), 815, 838–39. "Banker's Burial Service Monday," *RMN,* October 26, 1918. *Colorado Deluxe Supplement* (Chicago: S.J. Clarke Publishing Company, 1918), 33–35. Herbert O. Brayer, "Boom Town Banker," *The Westerners, Denver Posse, Brand Book* 3 (1947), 27–35.

11
Park County, Book F, 409–10. Stevens's also included mining deeds for 160 acres of placer claims for which he received $500, in addition to the $500 he received for the lode claims.

12
Park County, Book F, 411, 448.

13
Rossiter W. Raymond, *Statistics of Mines and Mining in the States and Territories West of the Rocky Mountains 1871* (Washington: GPO, 1873), 366.

Chapter 5

1
"The Fairplay Silver Mines," *DTB*, November 23, 1871. "The Mounts Lincoln and Bross Silver Mines," *RMN*, December 13, 1871. "Fairplay," *DTB*, December 13, 1871. "Out Look of Mining in Colorado for the Year 1872," *Georgetown Colorado Miner*, January 25, 1872.

2
"Park County Letter," *RMN*, July 30, 1872. Horace B. Patton et al., *Geology and Ore Deposits of the Alma District, Park County, Colorado*, Colorado State Geological Survey, *Bulletin 3* (Denver: Smith-Brooks Printing Company, 1912), 151–52. About 100 houses were built at Quartzville in 1872, plus shops, saloons, a storehouse for ore, and a steam crushing and sampling works built by J.B. Chaffee, Eben Smith, and Dr. Morrison; "Silver Mines of Lincoln and Bross Mountains," *RMN*, March 16, 1873.

3
"More of the Silver Deposits on Mts. Lincoln and Bross," *DTB*, December 18, 1871.

4
Out West 1 (April 13, 1872): 5.

5
"Jottings on the Hoof," *RMN*, March 9, 1872.

6
See Eleanor Bradley Peters, *Edward Dyer Peters (1849–1917)* (New York: Knickerbocker Press, 1918), 17–18. "Edward Dyer Peters," *EMJ* 103 (March 3, 1917), 380–81.

7
Taken from "Special Correspondence," *MRV* 1 (October 1872), 4–5. Rossiter W. Raymond, *Statistics of Mines and Mining in the States and Territories West of the Rocky Mountains 1872* (Washington: GPO, 1873), 296.

8
"Territorial Notes," *RMN*, March 23, 1872. "The Territories," *RMN*, April 30, 1872.

9
"The New Silver Region," *Denver Daily Times*, July 16, 1872.

10
"From Park County," *DT*, June 28, 1872. "From Mt. Lincoln District—More Rich Developments," *DT*, July 9, 1872. "Bross and Lincoln," *DT*, July 27, 1872. "Park County Letter," *RMN*, July 30, 1872. "Park County," *Out West* 1 (August 8, 1872): 7.

11
See "Special Correspondence," *MRV* 1 (October 1872): 4–5.

12
In his report for 1874 the U.S. Commissioner of Mining Statistics, Rossiter W. Raymond said of the mining on Bross

and Lincoln: "Detailed descriptions of their workings would sound like accounts of indiscriminate quarrying." Rossiter W. Raymond, *Statistics of Mines and Mining in the States and Territories West of the Rocky Mountains 1874* (Washington: GPO, 1875), 382.

13
"The New Silver Region," *Denver Daily Times,* July 16, 1872.

14
See either "From Mount Lincoln," *DT,* August 16, 1872; or "Mining Summary—Colorado," *Mining and Scientific Press* 25 (September 28, 1872): 197.

15
Hall, Brunk, and Peter K. Klinefelter discovered the Dolly Varden on August 1, 1872; Dolly Varden lode, Patent Case File 847, BLM Records Group 49, National Archives, Washington, D.C. They gave a warranty deed for the Dolly Varden, Kansas, Collins, and Red Rock to Henry R. Wolcott on September 30, 1872, for $2,000, but Wolcott gave the warranty deed back on November 4, 1872; Park County, Book 1, 168, 250.

16
"Park County," *Out West* 1 (October 10, 1872): 9. "Park County," *Out West* 1 (November 7, 1872): 7. "Mount Lincoln," *MRV* 1 (November 1872), Supplement, 2. "Park County," *Out West* 1 (November 14, 1872): 7. "Park County," *Out West* 1 (November 21, 1872): 7. "Mining Summary—Colorado," *Mining and Scientific Press* 25 (November 23, 1872): 325. "The Mount Lincoln Silver Mines," *Mining and Scientific Press* 25 (December 14, 1872): 370. "Fairplay Silver Mines," *DT,* May 31, 1873.

Chapter 6

1
Samuel Franklin Emmons, *Geology and Mining Industry of Leadville, Colorado,* USGS Monograph 12 (Washington: GPO, 1886), 107–23. Patton et al., *Geology and Ore Deposits,* 133–39. S.F. Emmons et al., *Geology and Ore Deposits of the Leadville Mining District, Colorado. USGS Professional Paper 148* (Washington: GPO, 1927), 27–33, 46–48. Quentin D. Singewald and B.S. Butler, "Suggestions for Prospecting in the Alma District, Colorado," *Colorado Scientific Society Proceedings* 13 (1933), 89–131.

2
See map in *MRV* 3 (September 1873): 4.

3
See "Official Reports, Park County, W.A. Smith, Assayer," *MRV* 1 (January 1873): 5.

4
Rossiter W. Raymond, "The Law of the Apex," *American Institute of Mining Engineers Transactions* 12 (1883–1884), 387–444.

5
"Mining Excitement at Fairplay," *DTB,* July 28, 1871.

6
"More of the Silver Deposits on Mts. Lincoln and Bross," *DTB,* December 18, 1871.

7
Curtis H. Lindley, *A Treatise of the American Law Relating to Mines and Mineral Lands* (San Francisco: Bancroft-Whitney Company, 1914), 1:666–75 and 2:982–83. Morrison, *Mining Rights in Colorado,* 115.

8

See Ronald W. Tank, *Legal Aspects of Geology* (New York: Plenum Press, 1983), 373. According to Maury Reiber, owner of the Russia Mine, "in the limestone replacement deposits, usually there was a connection between ore bodies. The old miners called them 'signboards,' [which] was a very small seam of ore sometimes only 1/16 or 1/8 inch wide. After cleaning out an 'ore pocket,' the miners would search the walls, back and floor for a 'signboard' which would lead them to the next ore deposit." Letter to author, November 2, 1999.

9

Park County, Book D, 150.

10

Park County, Book F, Wilson, 380, and Haines, 381. Curtis H. Lindley, *A Treatise of the American Law Relating to Mines and Mineral Lands*, 1: 646–47.

11

Don L. Griswold and Jean Harvey Griswold, *History of Leadville and Lake County, Colorado: From Mountain Solitude to Metropolis* (Denver: Colorado Historical Society and University Press of Colorado, 1996), 95, 1715–16. See also Raymond, "The Law of the Apex," 387–444.

12

"Park County Correspondence. Montgomery, Oct. 28," *Caribou Post*, November 4, 1871. Rossiter W. Raymond, *Statistics of Mines and Mining in the States and Territories West of the Rocky Mountains* (Washington: GPO, 1873), 366.

13

"Death of a Pioneer," *RMN,* December 11, 1893.

14

Joseph A. Thatcher Placer, mineral survey 75, file 13, 157A, and Hall and Morse Placer, mineral survey 89, file 13, 157B, BLM Records Group 49, National Archives, Washington, D.C.

15

Report of the Commissioner of the General Land-Office 1873 (Washington: GPO, 1873), 18.

16

James V. Dexter Papers, folder 106, CHS. This document is entitled "The William H. Stevens Circular published in 1873 or 1874."

17

Park County, Book B, 249–50, and Vol. 3, 352, 354.

18

Hall and Morse Placer, file 13157B, BLM Records Group 49, National Archives, Washington, D.C.

19

Ibid.

20

"An Important Mining Decision," *RMN,* March 22, 1874. "Mining Decisions and Legal Notes," *MRV* 4 (June 1874): 40–41. Park County, Vol. 5, 284. William K. Smith and James F. Flanagan, owners of the Eagle lode, quit-claim deeded the Eagle and other mining properties on Mount Lincoln to Joseph A.Thatcher. Thatcher held them in trust, whereupon the Alma Pool Association was organized, issuing $60,000 in stock (2,400 shares at $25 each). Smith owned $8,325 (333 shares); Flanagan owned $6,650 (266 shares); and William H. Stevens owned $5,400 (216 shares). James V. Dexter Papers, CHS.

Chapter 7

1
"The Ore Blockade," *MRV* 1 (October 1872): 1.

2
"What Are Our Resources?" *MRV* 1 (December 1872): 1–2.

3
"Foreign Reduction of Ores," *MRV* 2 (May 1873): 35.

4
Richard A. Ronzio, "Colorado Smelting and Reduction Works," *The Westerners, Denver Posse, Brand Book* 22 (1966): 109–45.

5
T. A. Rickard, *Pyrite Smelting* (New York: Engineering & Mining Journal, 1905), 11.

6
"Park County Letter," *RMN*, July 30, 1872.

7
Park County, Vol. 1, 412. "Fairplay Mining," *RMN*, December 13, 1872. "Park County," *RMN*, May 29, 1873.

8
Park County, Vol. 1, 327–30.

9
See Eleanor Bradley Peters, *Edward Dyer Peters (1849–1917)* (New York: Knickerbocker Press, 1918), 18–19.

10
See Fell, "The Boston and Colorado Smelting Company, 75–76, or "Park County," *RMN*, May 29, 1873.

11
"Territorial Notes," *RMN*, September 4, 1873.

Chapter 8

1
"Park County," *RMN*, March 11, 1873. "Park County," *RMN*, April 1, 1873. "From Park County," *RMN*, May 6, 1873. "Park County," *RMN*, May 29, 1873.

2
See "Silver Mines of Lincoln and Bross Mountains," *RMN*, March 16, 1873.

3
"From Park County," *RMN*, May 6, 1873.

4
"Park County," *MRV* 2 (March 1873): 2. "South Park and Arkansas Mines," *MRV* 2 (May 1873): 34.

5
Dolly Varden, Patent Case File 847, BLM Records Group 49, National Archives, Washington, D.C. "Park County," *MRV* 2 (June 1873), 45.

6
"The Hayden Expedition," *RMN*, July 22, 1873. Several members of the expedition spent the night of July 8, 1873, at the Present Help Mine, although for an unexplained reason they thought that they had stayed at the Montezuma Mine. A title search yields no information concerning the Montezuma. In July 1873 a building stood as living quarters for the miners at the Present Help. Following the night at the Present Help, William Henry Jackson took some photos from the summit of Mount Lincoln. His photo of the west ridge of Lincoln shows a mine building which can only be the Present Help, even though the photo contains the notation "Montezuma."

7

"Mt. Lincoln Mines," *DT,* August 30, 1873.

8

"Park County," *MRV* 2 (July 1873): 59. Present Help, Patent Case File 5346, BLM Records Group 49, National Archives, Washington, D.C.

9

"Fairplay Silver Mines," *DT,* May 31, 1877. "The Moose Mine," *MRV* 3 (September 1873): 10. "Park County," *MRV* 3 (December 1873): 48. "Mines and Mining," *DM,* August 24, 1873. "The Territorial Fair," *MRV* 3 (October 1873): 14–15. "The Fair," *RMN,* October 2, 1873. The Dolly Varden sent silver specimens assaying at $1,535.23 per ton.

10

"Mt. Lincoln Mines," *DT,* August 30, 1873.

11

"A great fire has been burning timber between Buckskin and Quartzville gulches for the past two weeks," noted the *Denver Mirror* ("Fairplay, Park Co.," July 13, 1873). See also *Hand-Book of Colorado* (Denver: J. A. Blake, 1872), 71–72. *Hand-Book of Colorado* (Denver: J. A. Blake, 1873), 44–45. *FF,* March 31, 1881. "The Burning of Fairplay," *DT,* September 29, 1873. "Territorial Jottings," *Colorado Springs Gazette,* October 4, 1873. "Fairplay, Park County," *DM,* October 5, 1873. *The Fairplay Sentinel* resumed publishing in December ("Fairplay, Park County," *DM,* December 21, 1873). It was not the first fire Fairplay had seen. On the night of July 31, 1872, the inhabitants had also been startled by a fire and feared it would destroy a large part of town, but there was no wind and there was a plentiful supply of water in the ditch nearby. "News of the Week," *Out West* 1 (August 8, 1872): 8. "Thirty Years Ago," *FF,* September 25, 1903.

Chapter 9

1

Charles W. Henderson, *Mining in Colorado. USGS Professional Paper 138* (Washington: GPO, 1926), 196. Clear Creek County, centered around Georgetown and Silver Plume, was the largest silver producer.

2

Edward D. Peters, "The Mount Lincoln Smelting Works at Dudley, Colorado," *EMJ* 17 (March 28, 1874): 194–95.

3

Fell, "The Boston and Colorado Smelting Company," 52–60. While Hill introduced the reverberatory to Colorado, it was already being used to smelt copper at other locations in the United States.

4

Park County, Vol. 3, 154, 255. Edward D. Peters, "The Mount Lincoln Smelting Works at Dudley, Colorado," *EMJ* 17 (March 28, 1874): 194–95. "What the Papers Are Saying," *DT,* March 7, 1874. Peters, *Edward Dyer Peters (1849–1917),* 21. Peters said, "I learned a lot about smelting here, but I am afraid the stockholders also gained a considerable amount of quite new experience."

5

Wolcott to Stevens, October 14, 1873, and Stevens to Wolcott, February 21, 1874, James V. Dexter Papers, folder 106, CHS.

6
Ibid.

7
Ibid.

8
"Does Mining Pay?" *DT,* January 19, 1874.

9
"Park County," *MRV* 4 (May 1874): 36. "Editorials," *MRV* 4 (June 1874): 42. "Park County," *MRV* 4 (June 1874): 45.

10
"Mining Matters," *DT,* March 11, 1874. "Fairplay," *RMN,* May 13, 1874. "Editorial," *MRV* 4 (August 1874): 60–61.

11
MRV 5 (November 1874): 31. *MRV* 5 (December 1874): 45.

12
Fairplay Sentinel, quoted in *MRV* 5 (December 1874): 45.

13
RMN, September 22, 1875. A letter to the editor stated that the Holland Smelter closed for lack of funds. The works had been erected on the credit system, and a large quantity of ores was donated. "It was not the fault of the fuel nor the furnace to which past failures are attributable, but to the incompetentcy soley [*sic*] of the person [Schafer] in charge of the enterprise." "Reduction of Colorado Ores: Why the Smelting Works at Holland Are Closed, etc.," *DM,* December 13, 1874. Schafer, replying to the *Fairplay Sentinel,* stated that he deserved the title "Dr."; doubters should contact either the Long Island Hospital Medical College or the War Department, for which he had served as a surgeon during the Civil War; *MRV* 5 (January 1875): 66.

14
MRV 4 (April 1874): 22–23. "Fairplay, Park County," *DM,* April 26, 1874. "Fairplay, Park County," *DM,* May 3, 1874. "Park County," *RMN,* May 10, 1874.

15
"Fairplay, Park County," *DM,* August 1, 1875.

16
"Editorials," *MRV* 4 (June 1874): 43. "Mount Lincoln Silver Mining District," *MRV* 6 (April 1875): 41–42. "Park County," *MRV* 5 (January 1875): 62. Henderson, *Mining in Colorado,* 196.

Chapter 10

1
"Mines and Mining," *DM,* August 24, 1873. "As things are here now, a poor man has no more chance than if he were in the desert of Sahara, for the capitalists have gobbled up everything in the way of the good mines on Mts. Lincoln and Bross, and outside of that there is nothing but low grade discovered yet."

2
Park County, Vol. 3, 206.

3
Park County, Vol. 5, 67. When the Alma Pool Association organized in July 1874, the remaining Park Pool Association properties were transferred to it. James V. Dexter Papers, CHS.

4
Charles S. Richardson Papers, Notebook 25, DPL. "That Musk Ox, etc.," *DM,* March 31, 1877. Richardson arrived in Denver on May 12, 1871. "Personal," *RMN,* May 13, 1871. He remained at Alma until at least 1885, where he

did some assaying and also had a "museum" of Park County ore specimens. He wrote a number of letters to the London *Mining Journal* about the minerals and mining in Park County. It is difficult to locate a copy of the *Mining Journal* for 1875–8, but letters 8–11 by Richardson appear in that journal as follows: 49 (September 13, 1879): 938; 50 (January 3, 1880): 20; 50 (April 10, 1880): 415–16; and 50 (October 9, 1880): 1160. Richardson also wrote a letter from Alma on January 1, 1885, describing the mines around Montgomery; "Montgomery Mines," *RMN*, January 6, 1885.

5

Charles S. Richardson Collection, Notebook 25.

6

Ibid. Richardson's report no longer exists, but several newspaper articles excerpted lengthy portions of it: *DM*, March 5, 1876; March 12, 1876; March 19, 1876. His survey notebooks also contain sketches of the mines on Bross and Lincoln.

7

"Park County," *MRV* 4 (March 1874): 12. "Gold and Silver," *DM*, March 22, 1874. "Park County," *MRV* 5 (November 1874): 32. "The Moose Mine," *RMN*, November 3, 1874. "The Precious Metals," *DM*, April 4, 1875, and April 18, 1875. "Colorado," *MRV* 6 (May 1875): 61. "The Precious Metals," *DM*, May 16, 1875. "Saunterings in South Park," *RMN*, November 9, 1875. "Park County Mines," *DM*, December 26, 1875.

8

"Our Mines," *DM*, January 23, 1876; February 6, 1876. "Park County,"

MRV 8 (March 6, 1876): 196. The Moose Mining Company owned the following claims: Moose (patent 546), Dwight (patent 573), Dudley (patent 574), Bross (patent 575), Gill (patent 576), Alps (patent 1358), Hoosier (patent 1359), Silver Star (patent 1369), Highland (patent 2912), Baker (patent 2913), Belle Gill (patent 2915), McNab (patent 3029), Emma (patent 3635), Capt. Plummer (patent 4081), Tunnel (patent 4114), and Julia (patent 4471).

9

See "Our Mines," *DM*, July 2, 1876; August 6, 1876; August 27, 1876; and September 17, 1876; October 8, 1876. "Fairplay, Park County," *DM*, December 9, 1876. Frank Fossett, *Colorado: A Historical, Descriptive and Statistical Work on the Rocky Mountain Gold and Silver Mining Region* (Denver: Daily Tribune Steam Printing House, 1876), 417.

10

"Fairplay, Park County," *DM*, March 7, 1875. "Gold and Silver," *DM*, May 24, 1874. "Park County Mines," *DM*, December 26, 1875. "Mining News," *EMJ* 23 (June 30, 1877): 459. "The Great Sale," *FF*, February 24, 1881. "Dolly Varden Property," *MR* 9 (March 12, 1881): 253.

11

Charles S. Richardson Collection, Notebook 25. Richardson's report on the Dolly Varden is excerpted in *DM*, February 20, 1876; May 14, 1876; September 10, 1876.

12

"New Corporations," *RMN*, May 5, 1877. The Hall and Brunk Silver Mining Company owned: Dolly Varden (patent 847), Hiawatha (patent 579), Tunnel

No. 1 (patent 1350), Tunnel No. 2 (patent 1349), Compromise (patent 1351), Jo Thatcher (patent 1352), Juniata (patent 1353), German (patent 1617), Polaris (patent 2808), Friday (patent 2809), Undercliff (patent 2810), and Iron Dyke (patent 2811). Deeds for the Hiawatha, Compromise, Juniata, and Jo Thatcher cannot be found in the Park County records. The German cost $2,000; Park County, Vol. 9, 146.

13
"Centennial," *RMN*, November 25, 1876. "Mines and Mining," *DM*, July 7, 1877. "The Centennial Awards to Colorado Mines," *RMN*, October 17, 1877. "Dolly Varden Property," *MR* 9 (March 12, 1881): 253. Carlyle Channing Davis, *Olden Times in Colorado* (Los Angeles: Phillips Publishing Company, 1916), 98.

14
Long Illness Ends in Death," *RMN*, December 2, 1905. "A. R. Meyer Is Dead," *Kansas City Times*, December 2, 1905.

15
L. A. Kent, *Leadville: The City, Mines and Bullion Product* (Denver: Daily Times Steam Printing House, 1880), 171–73. "Fairplay, Park County," *DM*, February 21, 1875. "Mount Lincoln Silver Mining District," *MRV* 6 (April 1875): 41–42. "Mill Notes," *MRV* 6 (July 1875): 107. "Fairplay, Park County," *DM*, September 12, 1875. "Park County," *MRV* 7 (September 20, 1875): 5. "Park County," *MRV* 7 (October 18, 1875): 38.

16
"Park County," *MRV* 7 (November 15, 1875): 70. "Smelting and Mill Notes," *MRV* 7 (December 13, 1875): 103.

"Park County Mines," *DM*, December 26, 1875. "Gilpin County," *MRV* 7 (January 10, 1876): 133. Henderson, *Mining in Colorado*, 196.

17
"Park County," *MRV* 7 (October 18, 1875): 38.

18
Daniel Plummer bought and later sold the Yankee Doodle Mine at Leadville for $300,000 before returning to Alma. He was always referred to as "Capt. Plummer" because he had been a captain in Company A, 27th Michigan Volunteer Infantry. He was mayor of Alma at one point, and he lived in Alma until May 1905, when his daughter Elizabeth accompanied him to Cedar Falls, Iowa. He died there on May 10, 1910, and is buried in the Greenwood Cemetery. *Park County Bulletin*, May 26, 1905. "Alma," *FF*, May 26, 1905. Joseph H. Myers moved from Fairplay to Denver in 1882. He became involved with the Little Hope claim in the Monarch District, Chaffee County, and, with several other men, put a title bond on the claim, and invested $10,000 in development. Then the original locator appeared, took control of the Little Hope, and after sinking the shaft further, found rich ore. Myers and his partners got a court order restoring the Little Hope to their control, but the owners who had given Myers and partners the title bond said they had not fulfilled the conditions of their bond. Myers and partners sued but lost the case on August 13, 1885. The *Fairplay Flume* wrote: "It is feared that Mr. Myers will be a heavy loser by reason of this, as the bulk of his small fortune was invested in this venture." Reportedly,

Meyers died on January 13, 1898, in Denver, but the only obituary that can be found is for a John H. Myers, age 61, an "expressman" whose wife was in the County Hospital. *FF*, October 12, 1882. Chaffee County District Court, Frank P. Davis et al. v. Alexander Hogue et al., Case 822, 1885. *Salida Mail*, June 26, 1885; August 14, 1885. "Mining Notes," *FF*, August 20, 1885. "He Died Suddenly," *DR*, January 14, 1898. Judson H. Dudley and the Moose Mine appear in a fictionalized story by Josephine Hart, "A Man of the West," *The Trail* 5 (July 1912), 5–20. Dudley's name is changed to "Darley," and the Moose Mine becomes the "Lynx Mine." The story portrays Dudley as regretting that he had sold the Moose. This has basis in fact: in 1899, when he was sixty-seven years old, Dudley organized the Moose Mining and Leasing Company to reopen the Moose. Work was "progressing night and day" when Dudley died suddenly in February 1900. His death took the steam out of the company, which did no more work and forfeited its lease. Richard B. Ware's connection with the Moose, other than that he was a partner when the company incorporated in April 1873, is unknown.

19
"Consolidated Montgomery and Buckskin Districts," *MRV* 6 (July 1875): 95. "Park County," *MRV* 7 (September 20, 1875): 5. "Mining Matters," *RMN*, January 10, 1877. "Mineral Production of Colorado for 1876," *EMJ* 23 (March 3, 1877): 137. "Fairplay, Park County," *DM*, December 9, 1876. "Our Mines," *DM*, January 20, 1877. "The Moose Mine," *RMN*, March 14, 1877. "A Rich Mineral County," *DM*, May 5, 1877.

"Fairplay, Park County," *DM*, May 5, 1877. "Park County," *RMN*, September 6, 1877. "Mines and Mining," *DM*, October 27, 1877. Davis, *Olden Times in Colorado*, 98.

Chapter 11

1
FF, July 31, 1884.

2
See "The Precious Metals," *DM*, December 6, 1874. "The Carbonate Mines of California Gulch," *EMJ* 25 (January 19, 1878): 40. "Leadville, Colo.—Colorado's Latest Excitement— A 'Camp' That Bids Fair to Become One of the Most Substantial in the West," *EMJ* 25 (March 30, 1878): 221. Stephen F. Smart, *Colorado Miner: An 1879 Guide to Leadville* (Kansas City, Mo.: Ramsey, Millett & Hudson, 1879), 5. Fossett, *Colorado: Its Gold and Silver Mines*, 404–8.

3
When Sullivan D. Breece died on November 17, 1877, he became the first internment in the Leadville Cemetery; Griswold and Griswold, *History of Leadville*, 262. Edward Blair, *Leadville: Colorado's Magic City* (Boulder: Pruett Publishing Company, 1980), 17.

4
Detroit News, January 25, 1910.

5
Information on California Gulch and early Leadville is drawn from Don L. Griswold and Jean Harvey Griswold, *The Carbonate Camp Called Leadville* (Denver: University of Denver Press, 1951). Blair, *Leadville: Colorado's Magic City*. Griswold and Griswold, *History of Leadville*.

6
Park County, Vol. 1, 154, 238, 245–46, 299, 333, 348; Vol. 5, 252; and Vol. 8, 221, 269, 306. Silver Gem (patent 1257) and Burnside (patent 2101).

7
"Leadville, Colorado's Latest Excitement," *EMJ* 25 (April 6, 1878), 239.

8
"An Important Purchase," *DM*, July 2, 1876.

9
"Letter from Leadville," *RMN*, May 14, 1878. "Leadville and the Iron Mine," *EMJ* 27 (February 15, 1879), 110–11.

10
See "Lake County, Colorado: A 'Smelter' Wanted," *EMJ* 23 (January 27, 1877): 55; also "Alma, Park County," *DM*, December 22, 1877. "Mines and Mining," *DM*, February 23, 1878.

11
"A Land of Silver and Lead," *NYT*, May 20, 1878. "Lodes and Laws at Leadville," *EMJ* 27 (April 26, 1879): 298. Griswold and Griswold, *History of Leadville*, 92, 175, 1715–16.

12
"Leadville's Columbus: A Talk with Mr. W. H. Stevens, Who Discovered the Great Treasure Vault," *DTB*, February 2, 1879.

13
Evening Chronicle, April 13, 1889, and *Carbonate Chronicle*, March 24, 1893, as quoted in Griswold and Griswold, *History of Leadville*, 1838, 1982.

14
FF, June 19, 1879. "Precious Metal Production of Colorado," *EMJ* 27 (January 11, 1879): 24.

15
See "Colorado's Latest Gift," *NYT*, February 16, 1879.

16
It took years for smelting in Colorado to become dominated by large smelters at Denver and Pueblo, but the fact that competition would force this to happen had been understood for some time. See "Transportation of Ores," *MRV* 2 (August 1873): 68. "The Refining Works at Black Hawk," *MRV* 3 (September 1873): 2. "Our Milling Capacity," *MRV* 3 (December 1873): 41–42. "Advance in Our Prices at Black Hawk," *EMJ* 23 (April 14, 1877): 231.

17
"Central Colorado," *RMN*, January 1, 1878. "The Alma Mines," *RMN*, May 17, 1878. "Park County," *EMJ* 26 (July 13, 1878): 27. "Point from Park," *RMN*, August 28, 1878. "Opening of Fairplay Branch," *RMN*, October 6, 1881.

Chapter 12

1
It is not known if John McNab invested in mining at Leadville. Andrew W. Gill did; see Griswold and Griswold, *History of Leadville*, 428, 488, 982. "Personals," *EMJ* 43 (April 9, 1887): 261, refers to a mining promotion in London by Andrew W. Gill, "Colorado promoter," and says: "Unless these gentlemen [Gill and S. V. Dorsey] have experienced an entire change of heart and practice, it may be taken for granted that their 'matured scheme' will leave the Britishers wiser but sadder men."

2
Many words have been written about investing in mining and mining fraud.

One contemporary example is the letter of T. E. Schwarz, M.E., "Precious Metal Mining Investments," *DR,* February 15, 1882. One of the best sources about mining, including mistakes in mining and fraud at the stock board, is Henry B. Clifford, *Rocks in the Road to Fortune; Or, the Unsound Side of Mining* (New York: Gotham Press, 1908). Also Joseph E. King, *A Mine to Make a Mine: Financing the Colorado Mining Industry, 1859–1902* (College Station, Tex.: Texas A&M University Press, 1977).

3
See "Gold and Silver Stocks," *EMJ* 25 (January 5, 1878): 10.

4
See "Investments in Mining," *MRV* 7 (February 7, 1876): 168–69.

5
Edwin P. Hoyt, *The Vanderbilts and Their Fortunes* (N.Y.: Doubleday and Company, Inc., 1962), 100, 111, 218. Also see Dorothy Kelly MacDowell, *Commodore Vanderbilt and His Family* (Hendersonville, N.C.: 1989), 29.

6
"Charles Mason Stead," *NYT,* October 14, 1926, 25. Thomas B. Corbett, *Colorado Directory of Mines* (Denver: Rocky Mountain News, 1879), 336–37.

7
"Park County," *RMN,* September 6, 1877. "Gold and Silver Stocks," *EMJ* 25 (January 12, 1878): 34.

8
"Stock Gambling," *EMJ* 25 (April 27, 1878): 287.

9
"Gold and Silver Stocks," *EMJ* 25 (May 11, 1878): 337.

10
Ibid. "Gold and Silver Stocks," *EMJ* 25 (June 8, 1878): 401.

11
"The Moose and Dolly Varden," *RMN,* March 29, 1878. "Colorado," *EMJ* 25 (April 27, 1878): 294. "The Alma Mines," *RMN,* May 17, 1878. "Park County," *EMJ* 25 (July 6, 1878): 9.

12
"Discovered Iron Ore in Michigan," *Detroit News,* December 30, 1903. "All Within the Lifetime of Jacob Houghton," *Detroit Tribune,* December 31, 1903.

13
"Gold and Silver Stocks," *EMJ* 27 (March 22, 1879): 210. "Aerial Mining," *FF,* March 13, 1879.

14
"The Moose Mining Company," *New York Daily Graphic,* February 21, 1879.

15
"Alma and the Mountains," *FF,* April 17, 1879. *FF,* May 15, 1879. "The Great Mines of Lincoln and Bross," *FF,* July 3, 1879. "Alma's Advance," *FF,* September 4, 1879. *FF,* September 18, 1879.

16
"Gold and Silver Stocks," *EMJ* 28 (November 22, 1879), 383.

17
See "Gold and Silver Stocks," *EMJ* 29 (January 24, 1880), 63.

18
See "Gold and Silver Stocks," *EMJ* 29 (January 31, 1880): 89.

19
"Mining Stocks," *MR* 7 (January 24, 1880): 92. "How It Is Regarded in Park County, Colorado," *MR* 7 (January 31,

1880): 98. "The Moose," *MR* 7 (February 7, 1880): 123. "Mining Stocks," *MR* 7 (February 7, 1880): 142.

20
"Moose Mine Shareholders," *MR* 7 (February 21, 1880): 172. "Mining Stocks," *MR* 7 (February 21, 1880): 189 and ibid. (February 28, 1880): 214. "The Moose Mortgage," *MR* 7 (March 13, 1880): 243.

21
FF, March 11, 1880. Initially, Edward M. Hawkins and John B. Bruner had gone to Leadville to start publishing a newspaper. The competition they saw there, however, encouraged them to try their luck elsewhere within Colorado's mining territories. They began publishing the *Fairplay Flume* in February 1879.

22
Report of the Board of Directors of the Moose Mining Company, March 1, 1880, DPL. A long section of the report appears in "The Moose Mine," *FF*, March 25, 1880.

23
Ibid. Crown (patent 7322, March 15, 1883), Emerald (patent 7323, March 15, 1883), Reliance (patent 7324, March 15, 1883), Argos (patent 9193, May 12, 1884), Addie, Mary and Dora (patent 17088, November 9, 1890).

24
"Park County 'Wild Cats,'" *FF*, January 22, 1880. *FF*, February 19, 1880. "Alma, Park County, Col., Feb. 16, 1880," *MR* 7 (February 28, 1880): 200–201. *FF*, March 11, 1880.

25
"Points from the Moose," *FF*, April 29, 1880.

26
"Over Capitalization of Mines," *FF*, April 29, 1880. The *New York Mining Directory*, published in March 1880, listed 577 mining companies, 144 of which were Colorado firms (25 percent). Four Mount Bross–Mount Lincoln mining companies appeared: on Lincoln, the Ford Consolidated, Musk Ox, and Russia; on Bross, the Moose. *New York Mining Directory* (New York: Hollister & Goddard, 1880), 17, 29, 33, 45.

27
"The Depression in Mining Stocks: The Cause and the Remedy," *EMJ* 29 (April 17, 1880): 268. "Gold and Silver Stocks," *EMJ* 30 (September 25, 1880): 211. "Mining Stocks," *MR* 7 (March 13, 1880): 261. "The Chrysolite," *MR* 7 (May 22, 1880): 482. One theory is that Andrew W. Gill and Company, brokers on Wall Street, started the raid on Little Pittsburgh stock; Griwsold and Griswold, *History of Leadville*, 488.

28
S. A. Nelson, *The Consolidated Stock Exchange of New York* (N.Y.: A. B. Benesch Co., 1907), 11. "George B. Satterlee," *New York Tribune*, September 20, 1903, 9. "George B. Satterlee," *NYT*, September 19, 1903, 7.

29
Park County, Vol. 16, 75–78. "New York February 1, 1881," *MR* 9 (February 5, 1881): 127. Robert A. Corregan and David F. Lingane, *Colorado Mining Directory* (Denver: Colorado Mining Directory Company, 1883), 548–49.

30
See "Gold and Silver Stocks," *EMJ* 31 (January 29, 1881): 82; vol. 32 (October 8, 1881): 243; (October 15,

1881): 258; (December 3, 1881): 379; vol. 33 (April 29, 1882): 225. "Mining Stocks," *MR* 10 (October 8, 1881): 357; (October 15, 1881): 380–81; vol. 11 (April 29, 1882): 404.

31
FF, October 7, 1880.

32
See "From the Snow-Clad Hills," *FF,* January 6, 1881. "The Moose," *MR* 9 (January 29, 1881): 108. *FF,* February 3, 1881. "Signs of Spring," *RMN,* February 22, 1881. "Mining Miscellany," *FF,* March 17, 1881. *FF,* May 19, 1881; July 14, 1881. "Park County," *EMJ* 32 (July 23, 1881): 59. "The Moose," *MR* 10 (July 30, 1881): 109. "Steady Producers," *FF,* August 18, 1881. *FF,* November 10, 1881; January 26, 1882; May 25, 1882. "Park County: One of the Old Reliable Districts," *Denver Republican Annual Supplement,* January 1, 1882, 44–45.

33
"Mining Stocks," *MR* 11 (May 13, 1882): 452.

34
See Park County, Vol. 19, 403, 630; also "Trouble in the Mining Exchange," *NYT,* May 14, 1882; "The Moose Conspiracy," *MR* 11 (May 13, 1882): 434–35; and "Mining Stocks," *MR* 11 (May 20, 1882): 476; vol. 12 (November 4, 1882): 449.

Chapter 13

1
"Mining," *RMN,* March 17, 1877. "The Moose and Dolly Varden," *RMN,* March 29, 1878. "Park County: One of the Old Reliable Districts," *Denver*

Republican Annual Supplement, January 1, 1882, 44–45.

2
The Park Pool Association properties in the area of the Hiawatha were: Hiawatha (patent 579), Compromise (patent 1351), Jo Thatcher (patent 1352), and Juniata (patent 1353). "The Precious Metals," *DM,* January 24, 1875, and March 28, 1875. "Mount Lincoln Silver Mining District," *MRV* 6 (April 1875): 41–42. "Park County," *MRV* 6 (August 1875): 134. "Park County," *MRV* 7 (October 18, 1875): 38. "Mining Intelligence," *RMN,* November 2, 1875. "Park County Mines," *DM,* December 26, 1875. Fossett, *Colorado: A Historical, Descriptive and Statistical Work,* 417.

3
"New Corporations," *RMN,* May 5, 1877. "The Dolly Varden Mine, Mt. Lincoln," *EMJ* 24 (August 28, 1877): 133. Corbett, *The Colorado Directory of Mines,* 330. Corregan and Lingane, *Colorado Mining Directory,* 536.

4
The Hall and Brunk Silver Mining Company prospectus is at DPL. "The Price of Ore," *RMN,* May 5, 1877. "The Alma Mines," *RMN,* May 17, 1878.

5
"New Corporations," *RMN,* May 5, 1877.

6
"Mount Bross Tunnel," *EMJ* 25 (June 22, 1878), 426–27.

7
"Mount Lincoln," *RMN,* June 8, 1878. "Sheriff's Sale," *FF,* April 21, 1881. One very negative assessment of the chances of success in the Mount Bross tunnel

appears in "Correspondence," *EMJ* 25 (June 8, 1878): 389–90. In 1903 J. E. Spurr of the USGS wrote: "Defunct tunnel companies, which started on nothing and ended in the same, are in many cases unheeded monuments to the falsity of the idea that veins will probably be found in depth, where there are no surface signs, even though the rock exposure may be excellent. Such companies start with the valiant determination of cross-cutting every vein in the country, and usually finish with a large dump." J. E. Spurr, "The Geologist in Matters of Practical Mining," *EMJ* 75 (April 11, 1903): 557.

8
"Great Mines of Lincoln and Bross," *FF*, July 3, 1879. "Colorado," *EMJ* 28 (October 4, 1879): 243–44. *FF*, November 6, 1879. "Alma Bulletin," *FF*, December 18, 1879; January 9, 1880. "Hostages to Fortune: The Wonderful Mines of the Mosquito District," *RMN*, March 11, 1880. Fossett, *Colorado: Its Gold and Silver Mines*, 505.

9
"An Exceptional Mine," *FF*, March 11, 1880. "The Dolly Varden," *RMN*, June 13, 1880.

10
Ibid. "Park County," *EMJ* 30 (July 17, 1880): 43.

11
"The Great Sale," *FF*, February 24, 1881. "Dolly Varden Property," *MR* 9 (March 12, 1881): 253. "Park County: One of the Old Reliable Districts," *Denver Republican Annual Supplement*, January 1, 1882, 44–45.

12
"Dolly Varden Mine," *MR* 9 (January 29, 1881): 108. *FF*, February 10, 1881.

13
J. Alden Smith, the first Territorial/State Geologist, received neither salary nor expenses and made his living by consulting for the mining industry. John W. Rold and Stephen D. Schwochow, *History of the Colorado Geological Survey (1872–1988)* (Denver: Colorado Geological Survey, 1989), 1.

14
"A Hit for the Hub," *RMN*, February 27, 1881. "The Dolly Varden," *DR*, February 28, 1881.

15
"A Big Mining Transaction," *RMN*, February 22, 1881. *FF*, March 3, 1881. Park County, Vol. 9, 617; Vol. 14, 463.

16
"Another Strike on the Dolly Varden," *FF*, April 14, 1881. "Official Letters: Dolly Varden," *EMJ* 31 (May 21, 1881): 358.

17
"Mining Intelligence," *DR*, July 2, 1881; July 8, 1881.

18
"From Breckenridge," *DR*, July 14, 1881. "The Boston Gold and Silver Mining Company," *MR* 10 (August 27, 1881): 196–97. "Park County," *EMJ* 32 (September 17, 1881): 189.

19
"Mining Intelligence," *DR*, August 24, 1881. "Receivers Sale," *FF*, December 8, 1881. Park County, Vol. 19, 404.

20
"Mine and Smelter," *RMN*, February 21, 1882. "Alma, Col., February 8, 1882," *MR* 11 (February 25, 1882): 177. " Mining Department," *DR*, March 1, 1882. "The Boston and Colorado Company," *FF*, May 4, 1882.

21

FF, September 21, 1882. Park County, Vol. 1, 103.

22

"Dolly Varden," *MR* 12 (December 30, 1882): 601. "Mineral Blossoms," *RMN,* March 18, 1884; *RMN,* June 3, 1884. *FF,* June 5, 1884. "Alma Local," *FF,* July 10, 1884. "Alma's Riches," *RMN,* September 2, 1884. *RMN,* September 10, 1884. "Alma's Wealth," *RMN,* October 4, 1884. "Park County," *EMJ* 38 (September 13, 1884): 180. "Park's Prosperity," *RMN,* February 22, 1885. "Mining Mention," *RMN,* August 21, 1885. *FF,* September 10, 1885. Park County, Vol. 26, 374, 377.

23

Park County, Vol. 63, 83, 87–88. The company's indebtedness also included a $10,000 promissory note signed by George W. Brunk on March 5, 1892. Park County, Vol. 48, 167.

Chapter 14

1

Hail and graupel are common at high altitude during the summer, but on the afternoon of August 11, 1881, a heavy black cloud settled over Mount Bross. For half an hour it was dark as night. By the time the cloud had risen and drifted off, it had left a solid six inches of hail all over the mountain. "A Fruitful Topic," *FF,* August 18, 1881.

2

For accounts of climbs to the top of Mount Lincoln, see Samuel Bowles, *Our New West* (Hartford, Conn.: Hartford Publishing Co., 1869), 144–48; Samuel Bowles, *The Switzerland of America: A Summer Vacation in the Parks and Mountains of Colorado* (Springfield, Mass.: Samuel Bowles & Company, 1869), 108–15; William H. Brewer, *Rocky Mountain Letters 1869* (Denver: Colorado Mountain Club, 1930), 27–30; George W. Pine, *Beyond the West* (Utica, N.Y.: T. J. Griffiths, 1871), 85–89; and John H. Tice, *Over the Plains, on the Mountains; Or, Kansas, Colorado, and the Rocky Mountains* (St. Louis: Industrial Age Printing Co., 1872), 195–97.

3

Grace Greenwood, "Notes of Travel," *NYT,* October 20, 1873.

4

Eliza Greatorex, *Summer Etchings in Colorado* (N.Y.: G. P. Putnam's Sons, 1873), 89.

5

Ibid, 89–90. Giraud Chester, *Embattled Maiden* (N.Y.: G. P. Putnam's Sons, 1951), 151–52.

6

Mark I. West, ed., *Westward to a High Mountain: The Colorado Writings of Helen Hunt Jackson* (Denver: CHS, 1994), 34.

7

Ibid, 35–36.

8

"Alma, Park County," *DM,* August 27, 1876.

9

Richardson Collection, Notebook 25, DPL.

10

"Alma, Park County," *DM,* February 6, 1876; March 12, 1876. During the winter of 1873–74, the miners at the Present Help produced 175 tons of ore; "Gold and Silver," *DM,* May 24, 1874.

11

"From Above," *FF,* November 18, 1880. *FF,* March 20, 1879.

12

"Dolly Varden," *RMN,* June 13, 1880.

13

"Western Items," *RMN,* May 3, 1877. "Alma and the Mountains," *FF,* April 17, 1879.

14

"The South Park Mines," *EMJ* 24 (December 29, 1877): 468–69.

15

Patton et al., *Geology and Ore Deposits,* 177–79.

16

"Fairplay, Park County," *DM,* August 27, 1876.

17

"Aerial Mining," *FF,* March 13, 1879.

18

"News from the Hills," *FF,* August 5, 1880. "From Another Correspondent," *FF,* August 5, 1880.

19

The 1880 census counted a total of 3,958 people in Park County, including 515 at Fairplay and 471 at Alma; "Census Figures," *FF,* July 15, 1880. Twelve votes were cast in the election of 1880 at Precinct 17, and twenty-three votes in the election of 1882. "Casting Up," *FF,* November 4, 1880. Fairplay cast 197 votes and Alma 355 votes in 1882; "The Vote by Precincts," *FF,* November 9, 1882.

20

FF, January 29, 1885.

21

FF, February 10, 1881.

Chapter 15

1

Park County, Vol. 54, 115–16.

2

"First Trace of F. H. Weeks," *NYT,* August 16, 1893.

3

Ludlow Patton was quite successful as a Wall Street banker and member of the New York Stock Exchange from 1851 to 1882, where he had a reputation for conservatism. He married Abby Hutchinson, one of the celebrated Hutchinson Family of singers, in 1849. Patton retired from business in 1873, and he and Abby traveled for about the next ten years. See "Ludlow Patton Dead," *NYT,* September 7, 1906. John Wallace Hutchinson, *Story of the Hutchinsons* (Boston: Lee and Shepard, 1896), 2: 273–75. Henry Whittemore, *The Founders and Builders of the Oranges* (Newark: L. J. Hardham, 1896), 269–74.

4

"Francis H. Weeks Deposed," *NYT,* May 2, 1893. "First Trace of F. H. Weeks," *NYT,* August 16, 1893. "Over Weeks Hangs a Sword," *NYT,* September 12, 1893. "The Sentence of Weeks," *NYT,* November 9, 1893.

5

Park County, Vol. 54, 115–16.

6

Colorado Division of Mines, Moose Mine, Mine Operator's Reports, 1980–81.

Bibliography

Manuscript and Archival Material

Chaffee County. District Court. Frank P. Davis et al. v Alexander Hogue et al. Case #822. 1885. Colorado State Archives, Denver.

Colorado. Division of Mines. Mine Operator's Reports. 1980-1981. Colorado State Archives, Denver.

Dexter, James V. Papers. Colorado Historical Society, Denver.

Hall and Brunk Silver Mining Company of Colorado. Prospectus. 1877. Denver Public Library, Western History Department.

Moose Mine. Report of Directors March 1, 1880. Denver Public Library, Western History Department.

Richardson, Charles S. Papers. Denver Public Library, Western History Department.

Van Wagenen, Theodore. Papers. Denver Public Library, Western History Department.

Public Documents

Emmons, Samuel Franklin. *Geology and Mining Industry of Leadville, Colorado. U.S.G.S. Monograph #12.* Washington: G.P.O., 1886.

————, J.D. Irving, and G.F. Loughlin. *Geology and Ore Deposits of the Leadville Mining District, Colorado. U.S.G.S. Professional Paper #148.* Washington: G.P.O., 1927.

Henderson, Charles W. *Mining in Colorado. U.S.G.S. Professional Paper #138.* Washington: G.P.O., 1926.

Park County. Pre-emption Book D.

————. Records.

Patton, Horace B., Arthur J. Hoskin, and G. Montague Butler. *Geology and Ore Deposits of the Alma District Park County, Colorado. Colorado Geological Survey Bulletin #3.* Denver, Colo.: State Printers, 1912.

Raymond, Rossiter W. *Statistics of Mines and Mining in the States and Territories West of the Rocky Mountains 1871.* Washington: G.P.O., 1873.

————. *Statistics of Mines and Mining in the States and Territories West of the Rocky Mountains 1872.* Washington: G.P.O., 1873.

————. *Statistics of Mines and Mining in the States and Territories West of the Rocky Mountains 1874*. Washington: G.P.O., 1875.

Rold, John W. and Stephen D. Schwochow. *History of the Colorado Geological Survey (1872-1988)*. Denver: Colorado Geological Survey, 1989.

United States. Census 1870 and Census 1880.

————. Commissioner of General Land Office. *Annual Reports 1873*.

————. National Archives. BLM Records. Record Group 49.

Books and Pamphlets

Blair, Edward. *Leadville: Colorado's Magic City*. Boulder: Pruett Publishing Co., 1980.

Bowles, Samuel. *Our New West*. Hartford, Conn.: Hartford Publishing Co., 1869.

————. *The Switzerland of America. A Summer Vacation in the Parks and Mountains of Colorado*. Springfield, Mass.: Samuel Bowles & Company, 1869.

Brewer, William H. *Rocky Mountain Letters 1869*. Denver, Colo.: Colorado Mountain Club, 1930.

Chester, Giraud. *Embattled Maiden*. New York: G.P. Putnam's Sons, 1951.

Clifford, Henry B. *Rocks in the Road to Fortune or the Unsound Side of Mining*. New York: Gotham Press, 1908.

Colorado. Deluxe Supplement. Chicago: S.J. Clarke Publishing Company, 1918.

Copp, Henry N. *Decisions of the Commissioner of the General Land Office and the Secretary of the Interior*. San Francisco: A.L. Bancroft & Co., 1874.

Corbett, Thomas B. *The Colorado Directory of Mines*. Denver, Colo.: Rocky Mountain News Printing Company, 1879.

Corregan, Robert A. and Lingane, David F. *Colorado Mining Directory*. Denver, Colo.: Colorado Mining Directory Company, 1883.

Daily Highest and Lowest Prices of Gold At New York for the Four Years From Jan. 1, 1862 to Jan. 1, 1866. New York: Commercial and Financial Chronicle, n.d.

Davis, Carlyle Channing. *Olden Times In Colorado*. Los Angeles, Calif.: Phillips Publishing Company, 1916.

Dill, R.G. *The Political Campaigns of Colorado*. Denver, Colo.: John Dove, 1895.

Encyclopedia of Biography of Colorado. Chicago: Century Publishing and Engraving Company, 1901.

Fell, James E. Jr. *Ores To Metals. The Rocky Mountain Smelting Industry.* Lincoln, Neb.: University of Nebraska Press, 1979.

Fossett, Frank. *Colorado: A Historical, Descriptive and Statistical Work on the Rocky Mountain Gold and Silver Mining Region.* Denver: Daily Tribune Steam Printing House, 1876.

———. *Colorado. Its Gold and Silver Mines, Farms and Stock Ranges, and Health and Pleasure Resorts.* New York: C.G. Crawford, 1879 and 1880.

Greatorex, Eliza. *Summer Etchings in Colorado.* New York: G.P. Putnam's Sons, 1873.

Griswold, Don L. and Jean Harvey. *The Carbonate Camp Called Leadville.* University of Denver Press, 1951.

———. *History of Leadville and Lake County, Colorado.* University Press of Colorado, 1996.

Hall, Frank. *History of the State of Colorado.* 4 vols. Chicago: Blakely Printing Co., 1889.

Hand-Book of Colorado. Denver: J.A. Blake, 1872 and 1873.

History of Colorado. 5 vols. Denver, Colo.: Linderman Co., Inc., 1927.

History of the Arkansas Valley. Chicago: O.L. Baskin & Co., 1881.

History of the City of Denver, Arapahoe County, and Colorado. Chicago: O.L. Baskin & Co., 1880.

Hoyt, Edwin P. *The Vanderbilts and Their Fortunes.* New York: Doubleday and Company, Inc., 1962.

Hutchinson, John Wallace. *Story of the Hutchinsons.* 2 vols. Boston: Lee and Shepard, 1896.

King, Joseph E. *A Mine to Make a Mine.* College Station: Texas A&M University Press, 1977.

Lindgren, Waldemar. *Mineral Deposits.* New York: McGraw-Hill, 1933.

MacDowell, Dorothy Kelly. *Commodore Vanderbilt and His Family.* Hendersonville, NC: The Author, 1989.

McConnell, Virginia. *Bayou Salado.* Denver, Colo.: Sage Books, 1966.

Manning, James H. *New York State Men.* Albany: Albany Argus Art Press, 1914.

Morrison, R.S. *Mining Rights in Colorado.* Denver: Chain & Hardy Publishers, 1881.

Nelson, S.A. *The Consolidated Stock Exchange of New York.* New York: A.B. Benesch Co., 1907.

New York Mining Directory. New York: Hollister & Goddard, 1880.

Peters, Eleanor Bradley. *Edward Dyer Peters (1849-1917)*. New York: Knickerbocker Press, 1918.

Pine, George W. *Beyond the West*. Utica, N.Y.: T.J. Griffiths, 1871.

Portrait and Biographical Record of the State of Colorado. Chicago: Chapman Publishing Company, 1899.

Rickard, T.A. *Pyrite Smelting*. New York: Engineering and Mining Journal, 1905.

Shamel, Charles H. *Mining, Mineral and Geological Law*. New York: Hill Publishing Co., 1907.

Smart, Stephen F. *Colorado Miner: An 1879 Guide To Colorado*. Kansas City, Mo.: Ramsey, Millett and Hudson, 1879.

Smiley, Jerome C. *History of Denver*. Denver, Colo.: J.H. Williamson & Company, 1903.

Stone, Wilbur Fisk. *History of Colorado*. Chicago: S.J. Clarke Publishing Company, 1918.

Tank, Ronald W. *Legal Aspects of Geology*. New York: Plenum Press, 1983.

Tice, John H. *Over the Plains, On the Mountains; Or, Kansas, Colorado, and the Rocky Mountains*. St. Louis: Industrial Age Printing Company, 1872.

West, Mark I. (ed.). *Westward to a High Mountain. The Colorado Writings of Helen Hunt Jackson*. Denver: Colorado Historical Society, 1994.

Whittemore, Henry. *The Founders and Builders of the Oranges*. Newark: L.J. Hardham, 1896.

Articles

Brayer, Herbert O. "Boom Town Banker." *The Westerners. Denver Posse. Brand Book*. 3 (1947): 27-60.

Morrison, Iva Evans, "William H. Stevens." *Colorado Magazine*. 21 (1944): 121-129.

Raymond, Rossiter W. "The Law of the Apex." *American Institute of Mining Engineers Transactions*. 12 (1883-1884): 387-444.

Ronzio, Richard A. "Colorado Smelting and Reduction Works." *The Westerners. Denver Posse. Brand Book*. 22 (1966): 109–146.

Singewald, Quentin D. and Butler, B.S. "Suggestions For Prospecting in the Alma District, Colorado." *Colorado Scientific Society Proceedings*. 13 (1933): 89-131.

Periodicals
Caribou Post November 4, 1871
Colorado Springs Gazette October 4, 1873
Colorado Tribune (Denver) 1870-1871
Denver Mirror 1873-1878
Denver Republican 1881-1882, 1898, 1900, 1904
Denver Times 1872-1874, 1879, 1903
Detroit Free Press March 8, 1901
Detroit News 1903, 1910
Engineering and Mining Journal 1874, 1877-1881, 1884, 1887, 1902-1903
Fairplay Flume 1879-1882, 1884-1885, 1899-1900, 1905
Georgetown Colorado Miner 1869-1872
Gloversville Daily Leader (N.Y.) October 28, 1899
Inter-Ocean v.1 (June 6, 1880)
Kansas City Times December 2, 1905
Mining and Scientific Press 1872
Mining Journal (London) 1879-1880
Mining Record (New York) 1880-1882
Mining Review 1872-1876
Monrovia Daily News (Calif.) June 9, 1917
New York Daily Graphic February 21, 1879
New York Times 1873, 1878-1879, 1882, 1893, 1903, 1906, 1926
New York Tribune September 20, 1903
Out West (Colorado Springs) 1872
Park County Bulletin (Alma) 1900-1901, 1905, 1910
Rocky Mountain Herald December 31, 1910
Rocky Mountain News 1861, 1865-1866, 1868-1869, 1871-1882, 1884-1885,
    1893, 1901, 1905, 1918
Salida Mail 1885
The Trail 1911-1912, 1917

Theses and Dissertations
Fell, James Edward Jr.. "Boston and Colorado Smelting Company: A Study
    in Western Industrialism." M.A. thesis, University of Colorado. 1972.

Interviews
Reiber, Maury. Owner of Russia Mine. Interviews with author. November
    12, 1996 and December 11, 1996.

Index

Abbe, William, 24
Allen, Daniel, 71-72, 76
Alma (Colo.), 43, 44, 48, 54, 61, 66, 68-69, 94, 98
Alps (mining claim), 20, 24

Baker Mine, 29, 46, 57
Baldwin, William, 26
Beeger, Hermann, 23
Bland-Allison Act, 100
Borders, John S., 99
Boston and Breckenridge Smelting Works, 87
Boston and Colorado Smelting Company, 74; Alma smelter, 43, 49, 55, 57, 60, 61-62, 69, 83; Black Hawk smelter, 3, 13-14, 41
Boston Gold and Silver Mining Company, 86, 87-88
Bradford, Horace, 80
Breece, Sullivan, 63-64
Bross — see Mount Bross
Bross (mining claim), 13, 16-19
Brunk, George, 20, 22, 28, 35, 36, 59-60, 82-90, 98; see also Hall and Brunk Silver Mining Company
Buckskin Joe, 4, 6, 7, 63

California Gulch, 63-64, 66, 68, 105
Chicago and New York Mining and Smelting Company, 53-54
Comstock (lode), 28, 67
Consolidated Montgomery District, 8-9, 25, 28-29, 36, 41, 43, 48, 49, 53, 56, 62

Dexter, James, 53
Dickinson, Anne, 92
Dolly Varden Mine, 28, 35, 46, 48, 53, 55, 59-62, 82-90, 92, 94-95, 97, 98, 99
Drummond, Willis, 36, 37
Dudley, Judson, 13-15, 16-19, 28, 41, 47, 62
Dudley (Colo.), 44, 54, 83
Dudley (mining claim), 13, 16-19
Dudley Works — see Mount Lincoln Smelting Works Company
Dwight (mining claim), 9-11, 14-15, 16-19, 66

Eagle (mining claim), 37

Fairplay, 6, 26, 27, 44, 48, 54, 69, 92
Fritz, John, 80

Gill, Andrew, 13-15, 16-19, 41, 62, 70
Gill (mining claim), 13, 16-19
Greenwood, Grace, 91-92
Greatorex, Eliza, 92

Hall, Assyria, 20, 22, 28, 35-37, 59-60, 82-90
Hall and Brunk Silver Mining Company, 60, 83
Harrison, Edwin, 66
Harrison Smelter, 66
Hayden Survey, 31, 46, 48
Henderson, Charles, 61
Hiawatha Mine, 20, 23, 29, 35, 46, 53, 82, 87
Hill, Charles, 72-73
Hill, Nathaniel P., 3, 23, 43, 50, 90

Holland, Charles, 53
Holland Smelter — see Chicago and New York Mining and Smelting Company
Hoosier Pass, 4
Hoosier (mining claim), 20, 24, 33
Horseshoe Mining District, 42, 50, 51
Houghton, Jacob, 73, 76-78, 80

Iron Mine, 66

Jackson, Helen Hunt, 92-93
Janes, Addison, 21
J.D. Dana Mine, 66

Lake County, 66
Leadville, 66-68, 71, 105
Lincoln — see Mount Lincoln
Lincoln Mine, 46

McNab, John, 19, 41, 62, 70
Meyer, August R., 60-62, 66, 69
mining law, 7-9, 16-19, 30-38; Mining Law of 1866, 16-17, 30, 32; Mining Law of 1870, 32; Mining Law of 1872, 33, 35, 37, 67; Placer Law of 1870, 33
Moffat, David, 72
Montgomery (Colo.), 4-6, 7, 25, 63
Montgomery mining district — see Consolidated Montgomery District
Moose Mine, 47, 66, 68, 94, 96-97; discovery of, 11, 13, 33; miners' quarters, 25, 29, 97-98; operations at, 27-28, 41, 46, 62, 72-74, 78, 80; patents on, 16-19; and smelting, 50-53, 55; transportation from, 14-15, 27, 47, 78, 95-97; yield of, 29, 47, 49, 57-59, 62, 74, 77
Moose Mining Company, 24, 62, 87; expansion of, 56-57; organization of, 16-19; stock price of, 70-77, 79, 80-81
Moose Silver Mining Company, 79-81, 102-104
Morse, Fred, 35-37
Mosquito (Colo.), 6, 63
Mosquito Range, 4, 7, 44, 63-65
Mount Bross: conditions on, 91-92, 93-94, 96-97; geology of, 30-34, 37, 65, 68, 77, 83-84, 105; mine development on, 7-9, 13, 17-18, 20-22, 24-27, 46-48, 56-62, 77; roads on, 46-47, 91-93, 95-97
Mount Bross Tunnel, 83-84
Mount Cameron, 9, 64
Mount Lincoln, 9; conditions on, 91-92; geology of, 30-34, 37, 65, 68, 105; mine development on, 21-22, 24, 27, 46-48; roads on, 46-47, 91-92
Mount Lincoln Smelting Works Company, 41-43, 44, 46, 49-55, 61-62, 69
Musgrove, William, 22
Myers, Joseph H., 6-15, 16-19, 20, 43, 62

New York Mining Stock Exchange, 70-72, 81, 84

Oro City, 63-64
Oro Mining Ditch and Fluming Company, 64-65

Panic of 1873, 71, 102

Park County: migration to, 4-6, 44, 68; mining activity in, 11, 20-21, 27, 49, 53, 60, 62, 69, 100

Park Pool Association, 28, 37-38, 43, 46, 48, 56, 60, 63, 68, 82; formation of, 23-24; and Boston and Colorado Smelter, 49-55

Patton, Ludlow, 103-104

Pease, Stephen, 63, 65

Peters, Edward D., 26, 31, 41, 50, 51

Phillips Mine, 4

Pitman, Fred, 99

placer mining, 1-2, 4, 6, 20, 32-35, 63-64; Placer Law of 1870, 33

Plummer, Daniel, 7-15, 16-19, 22, 43, 62

Present Help, 22, 47, 64, 91-94, 98

Quartzville, 25, 26, 41, 46, 93

Richardson, Charles, 57-60, 72, 84, 93-94

Rock Mine, 66

Rocky Mountain Mineral Concentration Company, 55

Russia Mine, 60, 62, 93, 94-95

Satterlee, George, 79-80

Sherman Act, 100

Silver Star (mining claim), 22, 24

smelting, 39-43, 49-55, 69

Stead, Charles, 71, 76

Stevens, William H., 22-24, 33, 35-36, 52-53, 56-57, 63, 64-68, 105

St. Louis Smelting Company, 60, 66, 83

Swansea (Wales), 15, 40, 42

Thatcher, Joseph, 23-24, 35, 53, 57

U.S. Land Office, 6, 18-19, 34-35, 37

Vanderbilt, Cornelius, 71

Vanderbilt, Ethelinda, 71

Ware, Richard B., 9-15, 16-19, 62

Weeks, Frances, 79-80, 81, 102-103

Wilson (mining claim), 21, 23, 33

Wolcott, Henry, 23, 35, 43, 52-53, 90

Wood, Alvinus, 36, 64-66, 68

LaVergne, TN USA
19 August 2010
193893LV00004B/4/A